Construction Maths

Advanced

Robert Neslen
Building Consultant, formerly Senior Lecturer
at North Lincolnshire College

A member of the Hodder Headline Group
LONDON · SYDNEY · AUCKLAND

First published in Great Britain in 1998 by Arnold,
a member of the Hodder Headline Group,
338 Euston Road, London NW1 3BH
http://www.arnoldpublishers.com

Whilst the advice and information in this book is believed to be true
and accurate at the date of going to press, neither the author nor the publisher
can accept any responsibility or liability for any errors or omissions
that may be made.

British Library Cataloguing in Publication Data
A catalogue record for this book is available from the British Library

ISBN 0 340 66237 9

Publisher: Eliane Wigzell
Production Editor: Rada Radojicic
Production Controller: Rose James
Cover Design: Andy McColm

Typeset by AFS Image Setters Ltd, Glasgow
Printed and bound in Great Britain by JW Arrowsmith Ltd, Bristol

Contents

chapter

1

Algebraic techniques

Outcomes

At the end of this chapter you should be able to:

- apply arithmetic progressions to various construction problems
- determine costs using simple interest techniques
- apply geometric progressions to various construction problems
- determine costs using compound interest techniques
- calculate mortgage repayments.

1.1 Linear equations

An algebraic equation states that one expression is equal to another, i.e.

$$4x = 12$$

Here we are saying that $4x$ is the same value or equal to 12. The x term is an unknown quantity. Because we are saying that the left-hand side of the equation is equal to the right-hand side, we can manipulate the equation how we like so long as we keep this basic truth, i.e. the left-hand side = the right-hand side of the equation.

Hence we could multiply the equation by 2 and, providing we do the same to both sides of the equation, the equality is maintained:

$$2(4x) = 2(12)$$

$$8x = 24$$

or we could divide both sides by 4:

$$4x \div 4 = 12 \div 4$$

$$x = 3$$

Equations with one unknown are often referred to as simple equations and providing the unknown is not raised to a power it is also a linear equation.

Linear equations can have more than one unknown and we shall be looking at these types of equation in more detail later on in the chapter.

To solve a simple equation we can use addition, subtraction, multiplication and division, always remembering that any mathematical operation that we perform on one side of the equation must be performed on the other side as well. Using these techniques we isolate the unknown without a numerical coefficient, so that we can say

unknown = ?

EXAMPLE 1.1

(a) $\qquad 6x = 42$

To isolate the unknown we divide both sides of the equation by 6:

$$6x \div 6 = 42 \div 6$$

$$x = 7$$

(b) $\qquad 4a = -20$

Dividing both sides by 4, we obtain

$$4a \div 4 = -20 \div 4$$

$$a = -5$$

(c) $\qquad 1.75q = 35$

$$1.75q \div 1.75 = 35 \div 1.75$$

$$q = 20$$

(d) $\qquad 1.5p = 6$

$$1.5p \div 1.5 = 6 \div 1.5$$

$$p = 4$$

(e) $\qquad 9 + x = 21$

To isolate x we must subtract 9 from each side:

$$9 - 9 + x = 21 - 9$$

$$x = 12$$

(f) $\qquad r - 2 = 9$

$$r - 2 + 2 = 9 + 2$$

$$r = 11$$

(g) $\qquad n \div 7 = 4$

Multiply throughout by 7 ro remove the fraction:

$$(n \div 7)(7) = (4)(7)$$

$$n = 28$$

(h) $117 \div x = 13$

Multiply throughout by x to remove the fraction:

$$(117 \div x)(x) = (13)(x)$$
$$117 = 13x$$

Divide both sides by 13 to isolate x:

$$9 = x \text{ or } x = 9$$

Some problems cannot always be solved by simple arithmetic and equations have to be formed to obtain a solution.

EXAMPLE 1.2

Thirty building operatives employed on a construction site work a 37 hour week. If the 30 operatives comprise tradesmen earning £4.75/hour and labourers earning £4.25/hour, how many tradesmen and how many labourers are there if the weekly wage bill is £5124.50?

Let the number of tradesmen be x.
Then the number of labourers $= 30 - x$.
In 1 week a tradesman earns $(4.75)(37) = £175.75$.
In 1 week a labourer earns $(4.25)(37) = £157.25$.
In 1 week x tradesmen earn £175.75x and labourers earn £157.25$(30 - x)$.

$$\therefore 175.75x + 157.25(30 - x) = 5124.5$$
$$175.75x + 4717.5 - 157.25x = 5124.5$$
$$175.75x - 157.25x = 5124.5 - 4717.5$$
$$18.50x = 407$$
$$x = 407 \div 18.50$$
$$x = 22$$

Therefore the number of tradesmen $= 22$ and the number of labourers $= 30 - 22 = 8$.

EXAMPLE 1.3

A rectangular school building is 15 m longer than it is wide. Given that its perimeter is 90 m, find its length and breadth.

Let the length $= x + 15$, and breadth $= x$. Then

$$2(x + 15) + 2x = 90$$
$$2x + 30 + 2x = 90$$
$$4x + 30 = 90$$
$$4x = 90 - 30 = 60$$
$$x = 60 \div 4 = 15$$

Therefore the breadth is 15 m and the length is 30 m.

EXAMPLE 1.4

A triangle has a perimeter of 700 mm. One side is 35 mm longer than the second and 12 mm shorter than the third. How long are the three sides?

Let side 1 $= x$, side 2 $= x - 35$ and side 3 $= x + 12$. Then

$$x + (x - 35) + (x + 12) = 700$$

$$3x - 23 = 700$$

$$3x = 700 + 23 = 723$$

$$x = 723 \div 3 = 241$$

Hence

$$\text{side 1} = 241 \text{ mm}$$

$$\text{side 2} = 241 - 35 = 206 \text{ mm}$$

$$\text{side 3} = 241 + 12 = 253 \text{ mm}$$

$$\text{check} = 700 \text{ mm}$$

See Section 1.6 for further application exercises.

1.2 Simultaneous linear equations

In the previous section we concentrated on the solution of equations with one unknown. We can now take this a step further and consider equations with more than one unknown. In order to do this we need to have more than one equation. A simple rule for this could be

- two unknowns – two equations
- three unknowns – three equations
- four unknowns – four equations, etc.

This simple rule is a good guide as to whether we can solve our equations.

Equations which can be solved like this are known as simultaneous equations. This means that there is a value for the unknowns however many equations we are considering. In the equations we are going to be looking at, the unknown terms will not be raised to a power greater than 1.

EXAMPLE 1.5

$$3x + 2y = 13$$

$$2x - y = 4$$

$x = 3$, $y = 2$ satisfy both equations, i.e. they have a simultaneous solution.

There are three methods in general use for the solving of simultaneous equations:

1. by graphical methods
2. by substitution
3. by elimination.

Method 1: Graphical

Since we are dealing with equations with power signs equal to 1 such as $2x + 3y = 8$, then they can be put in the form $y = mx + c$, the equation for a straight line graph.

For instance, $2y + 3x = 8$ can be rearranged to give

$$y = \frac{-3x + 8}{2}$$

$$y = -1.5x + 4 \ (y = mx + c)$$

If a pair of linear equations have a simultaneous solution, then if we were to plot these equations the two graph lines would intersect. The simultaneous solution is obtained from the point of intersection by reading the values off the horizontal and vertical axes (see Fig. 1.1).

Figure 1.1

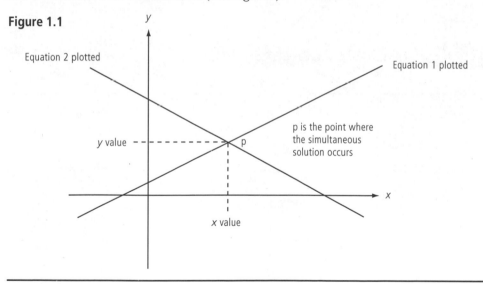

EXAMPLE 1.6

Solve by graphical methods the following pair of simultaneous equations:

$$x + y = 5$$
$$x - y = 1$$

by plotting their graphs over a range of $x = 0$ to $x = 6$

Since both equations will produce a straight line graph we need as a minimum the two end values for each line, but it is always a good idea to calculate a 'middle' value as a check against mistakes.

Equation $\quad x + y = 5$

when $x = 0 \quad y = 5$

$$x = 3 \qquad y = 2$$
$$x = 6 \qquad y = -1$$

Equation $\qquad x - y = 1$

when $x = 0 \qquad y = -1$
$$x = 3 \qquad y = 2$$
$$x = 6 \qquad y = 5$$

We can now plot the graphs of the two equations (see Fig. 1.2).

Figure 1.2

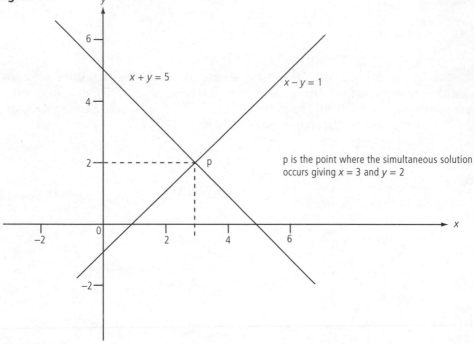

p is the point where the simultaneous solution occurs giving $x = 3$ and $y = 2$

To check the accuracy of our solution, substitute the values into the original equations:

$$3 + 2 = 5$$
$$3 - 2 = 1$$

EXAMPLE 1.7

Solve the given simultaneous equations by plotting their graphs:

$$x + 2y = 11$$
$$x + y = 6$$

Over the range of $x = 0$ to $x = 5$.

Equation $\qquad x + 2y = 11$

when $x = 0 \qquad y = 5.5$

$$x = 3 \qquad y = 4$$
$$x = 5 \qquad y = 3$$

Equation $\qquad x + y = 6$

$$\text{when } x = 0 \qquad y = 6$$
$$x = 3 \qquad y = 3$$
$$x = 5 \qquad y = 1$$

We can now plot the graphs of the two equations (see Fig. 1.3).

Figure 1.3

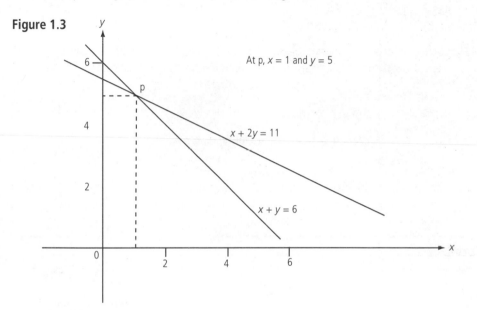

check: $\qquad 1 + (2)(5) = 11$
$$1 + 5 = 6$$

See Section 1.6 for further application exercises.

Method 2: Substitution

This method requires us to rearrange our equation to express one of the unknowns in terms of the other. Consider the following simple equation:

$$3x + 2y = 13$$

We can rearrange this equation to make x the subject:

$$x = \frac{13 - 2y}{3}$$

If $y = 1$ then $x = 3.67$; if $y = 2$ then $x = 3$, and so on.

It is this basic technique that we use for our substitution method as shown in the following examples.

EXAMPLE 1.8

Consider the equations

$$x - y = 8 \quad (1)$$
$$x + y = 12 \quad (2)$$

From (1), we obtain $x = 8 + y$. If we substitute this expression for x into equation (2), we obtain

$$8 + y + y = 12$$
$$2y = 12 - 8 = 4$$
$$y = 4 \div 2 = 2$$

Substitute this value for y into one of the equations. Since the values for x and y should be the same for both equations, it does not matter which one we substitute the value of y into.

$$x - 2 = 8$$
$$x = 8 + 2 = 10$$

As a check we should substitute both values into the equation (2):

$$10 + 2 = 12$$

which is correct. The simultaneous solution to both the equations is

$$x = 10 \quad \text{and} \quad y = 2$$

EXAMPLE 1.9

$$3a - b = 15 \quad (1)$$
$$a + b = 17 \quad (2)$$

From equation (1), $3a = 15 + b$. Therefore

$$a = \frac{15 + b}{3}$$

Substitute this value for a into (2):

$$\frac{15 + b}{3} + b = 17$$

To simplify, multiply throughout by 3:

$$15 + b + 3b = 51$$
$$4b = 51 - 15 = 36$$
$$b = 36 \div 4 = 9$$

Substitute this value for b into one of the equations

$$3a - 9 = 15$$
$$3a = 15 + 9 = 24$$
$$a = 24 \div 3 = 8$$

Substitute both values into the other equation:

$$8 + 9 = 17$$

Therefore the simultaneous solution is

$$a = 8 \quad \text{and} \quad b = 9$$

See Section 1.6 for further application exercises.

Method 3: Elimination

This method of solving simultaneous equations uses the technique of eliminating similar terms from each equation. This will leave us with a similar unknown term in each equation. The basic technique is to make the coefficient of the term to be eliminated the same in each equation. When this has been done, we can either add or subtract the equations to eliminate the term. Consider the following.

EXAMPLE 1.10

$$3x - y = 15 \quad (1)$$
$$x + y = 17 \quad (2)$$

In this example the y term has the same coefficient but the opposite sign. This suggests that we add the two equations together and so eliminate the y term:

$$3x - y = 15$$
$$\underline{x + y = 17}$$
$$4x = 32$$

If we now express this equation for x, we will have its value:

$$x = 32 \div 4$$
$$x = 8$$

Now substitute the value for x into one of the equations. I shall use equation (2) simply because it is the easiest:

$$x + y = 17$$
$$8 + y = 17$$
$$y = 17 - 8 = 9$$

To check our results are correct, substitute both values in equation (1):

$$3x - y = 15$$
$$(3)(8) - 9 = 15$$
$$24 - 9 = 15$$

The values for x and y are correct.

Let us go back to our original equation and eliminate the x term first:

$$3x - y = 15 \quad (1)$$

$$x + y = 17 \quad (2)$$

In order to make the coefficient of the x term in equation (2) the same as in equation (1), we must multiply by 3, i.e.

$$3x + 3y = 51 \quad (3)$$

Our two equations are now

$$3x - y = 15 \quad (1)$$

$$3x + 3y = 51 \quad (3)$$

The x terms now have the same numerical coefficient. Since the x terms have the same sign (+), in order to eliminate we must subtract:

$$3x - y = 15 \quad (1)$$

$$\underline{3x + 3y = 51 \quad (3)} \qquad \text{subtract (3) from (1)}$$

$$-4y = -36 \qquad \text{solving for } y$$

$$y = 9$$

We know this is the correct answer from our previous example.

These two examples demonstrate the basic techniques used to solve simultaneous equations by elimination:

1. Decide on which term to eliminate. If one of the terms has the same coefficient it would generally be the one to eliminate first, as in the first part of our example.
2. To eliminate the term we must look at the signs. If the signs are opposite then we add the equations together, as in the first part of our example. If the signs are the same then we must subtract, as in the second part of our example.

Remember the following rules when solving by elimination:

1. To make the coefficients the same we must multiply the entire equation, i.e.

$$x + y = 17 \quad \text{multiply by 3}$$

$$3x + 3y = 51$$

2. When we subtract we change the sign and add, i.e.

minus becomes plus
plus becomes minus

EXAMPLE 1.11

$$4x - 3y = 18 \quad (1)$$

$$x + 2y = -1 \quad (2)$$

To eliminate x, multiply (2) by 4:

$$4x - 3y = 18 \quad (1)$$
$$4x + 8y = -4 \quad (3)$$

Subtract (3) from (1):

$$4x - 3y = 18$$
$$\underline{4x + 8y = -4}$$
$$-11y = 22$$
$$y = 22 \div (-11) = -2$$

Substituting $y = -2$ into (2):

$$x + (2)(-2) = -1$$
$$x + (-4) = -1$$
$$x = -1 + 4 = 3$$

Check by substituting both values into (1):

$$4x - 3y = 18$$
$$(4)(3) - (3)(-2) = 18$$
$$12 + 6 = 18$$
$$\therefore x = 3 \quad y = -2$$

EXAMPLE 1.12

$$3x + 6y = 21 \quad (1)$$
$$4x - 5y = -11 \quad (2)$$

In this example I have decided to eliminate y simply because the signs are opposite. To make the coefficients the same I multiply equation (1) by 5 and equation (2) by 6, to give

$$15x + 30y = 105$$
$$\underline{24x - 30y = -66}$$

Adding, $39x = 39$

$$\therefore x = 1$$

Substitute into (1):

$$3(1) + 6y = 21$$
$$6y = 18$$
$$y = 3$$

Check by substituting into (2):

$$4(1) - 5(3) = -11$$
$$4 - 15 = -11$$
$$\therefore x = 1 \quad y = 3$$

So far our equations have had numerical coefficients that are whole numbers. This, of course, is not always the case and very often coefficients are in either fraction or decimal form. This makes no difference to solving the equation; the techniques are exactly the same. I think it is fair to say that the elimination method is probably the easiest to use in this situation.

EXAMPLE 1.13

Solve for x and y:

$$\frac{x}{6} + \frac{y}{8} = 3 \quad (1)$$

$$\frac{x}{3} - \frac{y}{4} = 0 \quad (2)$$

The first step is to remove the fractional form of the equations by multiplying each equation by the lowest common denominator. I think it is worth emphasising at this point that we can do more or less what we like to an equation providing we do the same thing to all the terms in the equation. It is of the utmost importance that we maintain the equality sign.

Starting with equation (1) we can multiply throughout by 24:

$$24\left(\frac{x}{6}\right) + 24\left(\frac{y}{8}\right) = 24(3) \quad (1)$$

This gives us

$$4x + 3y = 72 \quad (3)$$

If we now multiply equation (2) by 12

$$12\left(\frac{x}{3}\right) - 12\left(\frac{y}{4}\right) = 12(0)$$

This gives us

$$4x - 3y = 0 \quad (4)$$

Solving the equations

$$4x + 3y = 72$$

$$4x - 3y = 0$$

Adding the two equations together, we get

$$8x = 72$$

$$x = \frac{72}{8} = 9$$

Substituting the value into equation (3), we get

$$4(9) + 3y = 72$$

$$3y = 72 - 36$$

$$y = \frac{36}{3} = 12$$

Check by substituting both values into equation (4):

$$4(9) - 3(12) = 0$$

$$36 - 36 = 0$$

$$\therefore x = 9 \quad y = 12$$

EXAMPLE 1.14

Solve for a and b

$$2.5a - 1.5b = 2 \qquad (1)$$

$$0.75a + 3.75b = 51 \qquad (2)$$

With decimal fractions all we need to do is multiply the equations by multiples of 10, i.e. 10, 100, 1000, etc. to bring the coefficients to whole numbers.

Taking equation (1) and multiplying by 10, we get

$$25a - 15b = 20 \qquad (3)$$

and multiplying equation (2) by 100 we get

$$75a + 375b = 5100 \qquad (4)$$

Both these equations could be further simplified to make the coefficients more manageable by dividing (3) by 5 and (4) by 25:

$$25a - 15b = 20 \qquad \text{dividing by 5}$$

$$5a - 3b = 4$$

and $\qquad 75a + 375b = 5100 \qquad \text{dividing by 25}$

$$3a + 15b = 204$$

We can now solve these equations:

$$25a - 15b = 20$$

$$\underline{3a + 15b = 204}$$

$$28a = 224$$

$$a = 224 \div 28 = 8$$

Substitute this value into one of the equations:

$$5(8) - 3b = 4$$

$$40 - 3b = 4$$

$$-3b = 4 - 40 = -36$$

$$b = (-36) \div (-3) = 12$$

To check:

$$3(8) + 15(12) = 204$$

$$24 + 180 = 204$$

Sometimes the simultaneous equations are in reciprocal form, that is the alphabetical parts of the equations are below the line, i.e.

$$\frac{1}{x} \quad \text{or} \quad \frac{3}{b}$$

To solve these types of equations we can use either of the methods shown in the following two examples.

In the first example we follow the same method used in the previous examples, that is we make a term in each equation have the same value and eliminate.

EXAMPLE 1.15

Solve for m and n:

$$\frac{3}{m} + \frac{4}{n} = 24 \quad (1)$$

$$\frac{5}{m} - \frac{6}{n} = 2 \quad (2)$$

Multiplying (1) by 3 and (2) by 2, we get

$$\frac{9}{m} + \frac{12}{n} = 72$$

$$\frac{10}{m} - \frac{12}{n} = 4$$

Adding together gives us

$$\frac{19}{m} = 76$$

$$m = 19 \div 76 = \tfrac{1}{4}$$

Substitute this value into (1):

$$\frac{3}{\tfrac{1}{4}} + \frac{4}{n} = 24$$

$$12 + \frac{4}{n} = 24$$

$$\frac{4}{n} = 24 - 12 = 12$$

$$n = \frac{4}{12} = \frac{1}{3}$$

The alternative method involves the following substitution. Let $1/m = a$ and $1/n = b$. Then

$$\frac{3}{m} + \frac{4}{n} = 24$$

becomes

$$3a + 4b = 24 \quad (1)$$

and

$$\frac{5}{m} - \frac{6}{n} = 2$$

becomes

$$5a - 6b = 2 \quad (2)$$

Now solve as before:

$$3a + 4b = 24 \qquad \text{multiply by 5}$$
$$5a - 6b = 2 \qquad \text{multiply by 3}$$

$$15a + 20b = 120$$

$$\underline{15a - 18b = 6}$$

Subtracting, $38b = 114$

$$b = 114 \div 38 = 3$$

Substituting this value into (1), we get

$$3a + 4(3) = 24$$

$$3a = 24 - 12 = 12$$

$$a = 12 \div 3 = 4$$

But $a = 1/m$ and $b = 1/n$.

$$\therefore 4 = \frac{1}{m} \quad \text{and} \quad 3 = \frac{1}{n}$$

Transposing these equations we get

$$m = 1/4 \quad \text{and} \quad n = 1/3$$

Simultaneous equations

When solving simultaneous equations there are three steps to follow:

1. decide on the equations to represent the known details of the problem;
2. solve the equations choosing the method of solution you consider to be the most appropriate;
3. check the solutions.

EXAMPLE 1.16

In one full working week, i.e. 5 days, a gang of eight bricklayers and five apprentices can lay 27 200 bricks. Another gang of six bricklayers and four apprentices lay 20 800 bricks in the same time. Find the number of bricks laid per hour by

(a) a bricklayer
(b) an apprentice

given that one working day = 8 hours.

Let the number of bricks laid by a bricklayer = b and the number by an apprentice = a.

Then

$$8b + 5a = 27\,200 \quad (1)$$

and

$$6b + 4a = 20\,800 \quad (2)$$

These equations represent the output per week.

To solve, multiply (1) by 4 and (2) by 5 to give

$$32b + 20a = 108\,800$$

$$30b + 20a = 104\,000$$

Subtracting we get

$$2b = 4800$$

$$b = 2400 \text{ bricks/week}$$

Substituting back into (1)

$$8(2400) + 5a = 27\,200$$

$$19\,200 + 5a = 27\,200$$

$$5a = 27\,200 - 19\,200 = 8000$$

$$a = 8000 \div 5 = 1600$$

Check:

$$6(2400) + 4(1600) = 20\,800$$

$$20\,800 = 20\,800$$

These are the weekly totals per bricklayer and apprentice. Therefore divide by the number of hours in a week:

bricklayer: $2400 \div 40 = 60$ bricks/hour

apprentice: $1600 \div 40 = 40$ bricks/hour

EXAMPLE 1.17

A series of experiments carried out on a hoisting winch provide the formula of $E = aL + b$ where L represents the load and E the effort required to raise it. Two of the experiments used to obtain the formula were:

$$E = 8 \quad L = 23$$

$$E = 14 \quad L = 92$$

Determine the values of a and b from the experimental data.

Substituting the experimental data into the formula, we get

$$23a + b = 8 \quad (1)$$

$$92a + b = 14 \quad (2)$$

Multiplying (1) by 4 gives

$$92a + 4b = 32 \quad (3)$$
$$92a + b = 14 \quad (2)$$

Subtracting (2) from (3) gives

$$3b = 18$$
$$b = 18 \div 3 = 6$$

Substituting into (1):

$$23a + 6 = 8$$
$$23a = 8 - 6 = 2$$
$$a = 2 \div 23$$

Check by substituting into (2):

$$92(2/23) + 6 = 14$$
$$8 + 6 = 14$$
$$\therefore a = 2/23 \quad \text{and} \quad b = 6$$

See Section 1.6 for further application exercise.

1.3 Quadratic equations

A quadratic equation is an equation which contains an unknown quantity in the second degree, i.e. a quantity raised to the power of 2 such as x^2. It may also contain the first power of a quantity, i.e. $x^2 + x$, and it may also contain a numerical constant, i.e. $x^2 + x + 10$.

Further examples of quadratic equations are

$$x^2 = 36$$
$$6x^2 = 24$$
$$6x^2 + x = 26$$
$$6x^2 + x - 6 = 20$$
$$x^2 = -6x$$

The general form of a quadratic equation can be written as

$$ax^2 + bx + c = 0$$

where a, b, and c are constant numbers.

ALL QUADRATICS SHOULD BE SIMPLIFIED TO THIS FORM READY FOR SOLVING.

When solving quadratic equations we have two values of x which are called the roots of the equation. The reason for this is fairly straightforward. If we take the simple quadratic

$$x^2 = 36$$

we can see by observation that the value of x is equal to 6. It is also equal to –6, since a minus quantity multiplied by a minus quantity gives a positive quantity. Going back to our equation again, we can say that if

$$x^2 = 36$$

then

$$x = \pm 6$$

There are four methods we can use to solve quadratic equations

1. graphical (probably the easiest)
2. factorisation
3. completing the square
4. formula (the most generally used since it suits all situations).

Graphical method

We are looking to solve the equation of the form

$$ax^2 + bx + c = 0$$

that is to find the roots of the equation. These occur where the curve of the graph passes through the x axis.

EXAMPLE 1.18

Solve the equation $y = 2x^2 - 3x - 8$ over the range of values $x = -3$ to $x = 4$.

Before plotting the graph we have to calculate the value of y over the given range of values. The easiest way to do this and reduce the chances of error is to construct a box as shown below and work through it methodically:

when	$x =$	-3	-2	-1	0	1	2	3	4
	$2x^2 =$	18	8	2	0	2	8	18	32
	$-3x =$	9	6	3	0	-3	-6	-9	-12
	$-8 =$	-8	-8	-8	-8	-8	-8	-8	-8
then	$y =$	19	6	-3	-8	-9	-6	1	12

The equation is written vertically and for each value of x the y value is calculated, so that we can say

when $x = -3$ $y = 19$

$x = -2$ $y = 6$ etc.

These values are now the x and y coordinates of our graph (see Fig. 1.4).

The solution to our equation lies where the graph of $y = 2x^2 - 3x - 8$ crosses the x axis at points p and q. Reading off these values we find that x is equal to -1.4 and 2.9, the roots of our equation. This can be checked by substituting these values back into our equation:

when $x = -1.4$ $y = 0$ approximately

and when $x = 2.9$ $y = 0$ approximately

Figure 1.4

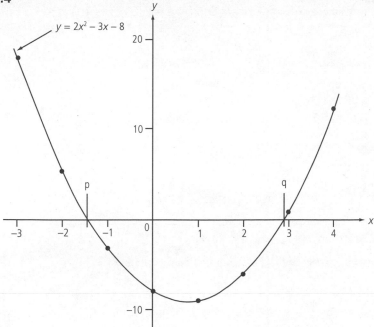

EXAMPLE 1.19

Plot the graph of $2x^2 - 5x + 2$ over the range of values $x = -1$ to $x = 3$.

When	$x =$	-1	0	1	2	3
	$2x^2 =$	2	0	2	8	18
	$-5x =$	5	0	-5	-10	-15
	$+2 =$	2	2	2	2	2
then	$y =$	9	2	-1	0	5

Figure 1.5 shows these coordinates for x and y plotted, and from the graph we can see that when $y = 0$, $x = 0.5$ and 2.

Solving quadratics by graphical methods is a fairly simple procedure but does rely on the plotting of an adequate number of coordinated points to allow us to achieve an acceptable accuracy in obtaining the roots.

It is not always necessary to plot a new graph for every equation providing the second degree term is the same.

EXAMPLE 1.20

Plot the graph of $3x^2$ and using this graph solve

(a) $3x^2 - 6$

(b) $3x^2 + 2x - 6$

(c) $3x^2 - x + 16$

Figure 1.6 shows the graph of $3x^2$.

Figure 1.5

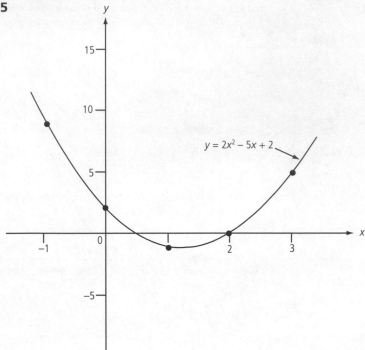

$y = 2x^2 - 5x + 2$

Figure 1.6

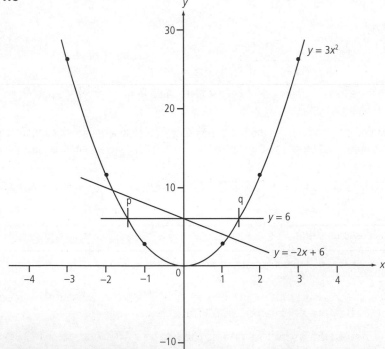

$y = 3x^2$

$y = 6$

$y = -2x + 6$

The equation $3x^2 - 6 = 0$ can be transposed to give $y = 3x^2 = 6$. This means it can be represented by the curve $y = 3x^2$ and the straight line $y = 6$. It is at the intersection of these two equations that the roots of the quadratic $3x^2 = 6$ lie:

$$y = 3x^2 = 6$$

that is

$$3x^2 - 6 = 0$$

These roots are shown at p and q giving the approximate values of $-1\cdot4$ and $1\cdot4$.

$3x^2 + 2x - 6$ can be solved in a similar manner. Firstly rearrange the equation to

$$3x^2 = -2x + 6$$

Now plot equation $-2x + 6$ on our graph of $3x^2$ to give the approximate values of $-1\cdot8$ and $1\cdot1$. The approximations occur simply because of the plotting inaccuracies.

We can use this technique to solve simultaneous linear and quadratic equations.

EXAMPLE 1.21

Find the simultaneous values of x and y for $y = 2x^2 - x - 1$ and $y = x + 2$ over the range of $x = -2$ to $x = 3$.

For there to be a simultaneous solution to the two equations, it follows that at some point $y = 2x^2 - x - 1$ must equal $y = x + 2$, i.e.

$$2x^2 - x - 1 = x + 2$$

If we now plot the two equations then the points of intersection will give us the solutions that we are looking for:

$$y = 2x^2 - x - 1$$

when	$x =$	-2	-1	0	1	2	3
	$2x^2 =$	8	2	0	2	8	18
	$-x =$	2	1	0	-1	-2	-3
	$-1 =$	-1	-1	-1	-1	-1	-1
then	$y =$	9	2	-1	0	5	14

$$y = x + 2$$

when $\quad x = -2 \quad y = 0$

$\qquad\quad\; x = 0 \quad\; y = 2$

$\qquad\quad\; x = 3 \quad\; y = 5$

The approximate solution to our simultaneous equations are $x = -0\cdot8$ and $x = 1\cdot8$ (see Fig. 1.7).

Factorisation

If a quadratic equation can be factorised, it is often the easiest way of solving it. We know the general form of a quadratic is $ax^2 + bx + c = 0$.

If this can be written as factors taking the form of $(x - a)(x - b) = 0$, we can find the roots of the equation.

Figure 1.7

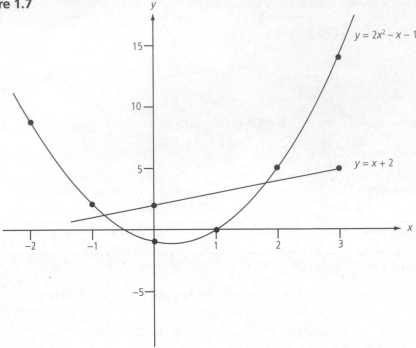

In the equation $(x - p)(x - q) = 0$, it is necessary to find the values of x which make the product zero. It is worth remembering that if the product of two quantities is zero then either or both of the quantities is equal to zero, i.e.

$$x - p = 0 \qquad \therefore x = p$$

or

$$x - q = 0 \qquad \therefore x = q$$

If the equation $ax^2 + bx + c = 0$ can be divided into its factors then three rules must be obeyed.

Let us assume that the factors to the general form of a quadratic $ax^2 + bx + c = 0$ are

$$(rx + p)(sx + q)$$

Then:

1. $rs = a$

2. $pq = c$

3. $rq + ps = b$

Consider the equation $3x^2 - 5x - 2 = 0$. Then, using the above rules

$$rs = a \quad \text{i.e. } 3(1) = 3$$

$$pq = c \quad \text{i.e.} -2(1) = -2 \text{ or } 2(-1) = -2$$

$rq + ps = b$ i.e. $rq = 3(-2) = -6$ and $ps = 1(1) = 1$

$rq + ps = -6 + 1 = -5$

Thus

$$(3x + 1)(x - 2) = 3x^2 - 5x - 2 = 0$$

$$\therefore\ 3x + 1 = 0 \quad \text{where } x = -1/3$$

$$\text{or } x - 2 = 0 \quad \text{where } x = 2$$

This can be checked by substituting the values into the original equation.

Very often factorisation of a quadratic can be solved by observation, but if the factors are not readily seen we must make use of the three rules.

EXAMPLE 1.22

Factorise $x^2 - 16 = 0$.

By observation $(x + 4)(x - 4) = 0$

where $x - 4 = 0$ $\therefore x = 4$

or $x + 4 = 0$ $\therefore x = -4$

EXAMPLE 1.23

Factorise $x^2 + x - 56 = 0$.

With this equation we are looking for two numbers whose product is -56 and whose summation is 1

$$-56 = -1(56) \quad \text{or} \quad -56(1)$$
$$= -2(28) \quad \text{or} \quad -28(2)$$
$$= -4(14) \quad \text{or} \quad -14(4)$$
$$= -7(8) \quad \text{or} \quad -8(7)$$

Now, the only two numbers that have a summation difference of 1 are -7 and 8; therefore our factors are

$(x + 8)(x - 7) = 0$

where $x + 8 = 0$ giving $x = -8$

or $x - 7 = 0$ giving $x = 7$

EXAMPLE 1.24

Factorise $15x^2 - 77x + 10 = 0$.

With equations like this it is not always possible to see the factors, so let us go back to our three rules. If the factors are $(rx + a)(sx + b)$ then we can write down the factors of a, b and c in the form of a table. We know then that the factors we are looking for are in the table:

r	×	s	p	×	q
1		15	1		10
3		5	2		5
−1		−15	10		1
−3		−5	5		2
15		1	−1		−10
5		3	−2		−5
−15		−1	−10		−1
−5		−3	−5		−2

It is necessary to realise that the order and signs are important. Now we have all the possible factors to give 15 and 10, we need to refine our table to find the two pairs of numbers that will give us 77. Since 77 is a fairly large number and 15 is the coefficient of x^2, this would suggest a good starting point for finding the factors more or less by trial and error. As shown on the table the two pairs of numbers that we are looking for are −15 and −1; 2 and 5.

Therefore our factors are $(-15x + 2)(-x + 5) = 0$. Then

$$-15x + 2 = 0$$

where

$$-15x = -2$$

and

$$x = -2 \div (-15) = 2/15$$

Also, $-x + 5 = 0$

$$-x = -5$$

$$x = -5 \div (-1) = 5$$

The roots of the equation are $x = 5$ and $x = 2/15$.

EXAMPLE 1.25

Factorise $7x^2 - 27x - 4 = 0$.

Setting up our table:

r	s	p	q
7	1	4	−1
1	7	1	−4
−7	−1	2	−2
−1	−7	−4	1
		−1	4
		−2	2

$$(7x+1)(x-4) = 0$$
$$7x+1 = 0$$
$$7x = -1$$
$$x = -1/7$$

and

$$x-4 = 0$$
$$x = 4$$

The roots of the equation are $-1/7$ and 4.

Completing the square

Before looking in more detail at the completing the square method, let us consider three special expressions that occur quite frequently in algebra:

1. $(a + b)(a + b) = (a + b)^2$

2. $(a - b)(a - b) = (a - b)^2$

3. $(a + b)(a - b) = a^2 - b^2$

If we expand (1) and (2) then we get

$$(a + b)(a + b) = a^2 + 2ab + b^2$$
$$(a - b)(a - b) = a^2 - 2ab + b^2$$

These trinomials are known as perfect squares and to recognise a trinomial as a perfect square it must conform to the following conditions:

- 1st term is squared

- 2nd term is equal to twice the product of the two terms

- 3rd term is squared.

It is on these conditions that we will be concentrating to complete the square. Before we do that, a quick word about $a^2 - b^2$. This is known as the difference between two squares. The three expressions shown are often referred to as identities because any value can be substituted for the letters. They represent 'patterns' that frequently occur in mathematics.

EXAMPLE 1.26

Write down the squares of

1. $(x + 3)^2 = x^2 + 6x + 9$

2. $(x - 2)^2 = x^2 - 4x + 4$

3. $(x + 5)^2 = x^2 + 10x + 25$

4. $(2x + y)^2 = 4x^2 + 4xy + y^2$

5. $(x + 0.5)^2 = x^2 + x + 0.25$

As you can see, if we know the pattern we can quickly solve our problem.

What must be added to $x^2 + 6x$ to make the trinomial a perfect square? That is

$$(x + ?)(x + ?) = (x + ?)^2$$

Now from our conditions we know that $6x$ represents twice the product so that the actual product is $3x$. Therefore the two terms required to make a perfect square are x and 3, to give $(x + 3)^2$:

$$(x + 3)^2 = x^2 + 6x + 9$$

We can demonstrate this geometrically as follows:

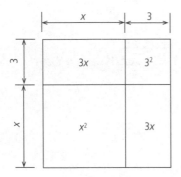

We say we are completing the square by 'adding the square of half the coefficient of x' and the square root of $x^2 + 6x + 9 = \pm(x + 3)$.

Remember there are two roots.

EXAMPLE 1.27

Give the two roots of the expression $x^2 - 10x$.

You will find it most useful to follow a logical sequence of steps as suggested below:

1. coefficient of x is -10

2. half the coefficient of x is -5

3. square (2): $(-5)^2 = 25$

4. add 25 to $x^2 - 10x$ to produce the perfect square

5. the square root of $x^2 - 10x + 25$ is $\pm(x + 5)$.

Therefore the roots are $(x - 5)$ and $(5 - x)$.

So far our method has only been applied to equations where the coefficient of x^2 is unity, i.e. 1.

When we consider equations where the coefficient of x^2 is not unity, we must adjust our equation so that it is unity and positive.

EXAMPLE 1.28

Give the two roots of $4x^2 - 20x$. Again follow a logical sequence to try and reduce the chances of making mistakes. The following is simply the previous sequence extended slightly:

1. remove the coefficient of x^2 as a common factor of the equation, i.e. $4(x^2 - 5x)$;

2. make the bracket a perfect square by adding the square of the coefficient of x, i.e. $(-5/2)^2$

   $4(x - 5x + (-5/2)^2) = 4x^2 - 20x + 25$;

3. the square root of $4x^2 - 20x + 25 = \pm(2x - 5)$;

Therefore the roots are $(2x - 5)$ and $(5 - 2x)$.

EXAMPLE 1.29

Find the roots of $5x^2 + 14x = 55$ by completing the square.

With this equation we have a numerical value on the right-hand side of the equation as well. The key point here is that whatever we do to one side of the equation we must do to the other.

1. make the coefficient of x^2 unity

2. $x^2 + \left(\dfrac{14}{5}\right)x + \left(\dfrac{14}{10}\right)^2 = 11 + \left(\dfrac{14}{10}\right)^2$

3. $(x + 1\cdot4)^2 = 11 + 1\cdot4^2$

4. $x + 1\cdot4 = \pm\sqrt{11 + 1\cdot4^2} = \pm\sqrt{12.96}$

   $x = \pm\sqrt{(12.96)} - 1\cdot4$

   $x = \pm3.6 - 1.4$

Therefore the roots of the equation $= 2\cdot2$ and -5.

SUMMARY OF COMPLETING THE SQUARE

1. Simplify the equation by moving the constant term, i.e. the numerical value, to the right-hand side of the equation so that only the x^2 and the x terms are on the left-hand side of the equation.

2. Make the coefficient of x^2 unity and positive by dividing throughout the equation by the value of the numerical coefficient including the sign.

3. Add the square of half the coefficient of x to both sides of the equation.

4. Factorise the left-hand side of the equation.

5. Take the square root of both sides of the equation.

6. Solve the resulting equation and remember that there are two roots.

EXAMPLE 1.30

Find the roots of $12x^2 = 29x - 14$ by completing the square correct to decimal places.

1. $12x^2 - 29x = -14$

2. $x^2 - \dfrac{29}{12}x = -\dfrac{14}{12}$

3. $x^2 - \dfrac{29}{12}x + \left(\dfrac{-29}{24}\right)^2 = -\dfrac{14}{12} + \left(\dfrac{-29}{24}\right)^2$

4. $\left(x - \dfrac{29}{24}\right)^2 = -1\cdot17 + 1.46$

$$= 0.29$$

5. $x - 1\cdot21 = \pm\sqrt{0\cdot29}$

6. $x = \pm\sqrt{0\cdot29} + 1.21$

$x = \pm 0\cdot54 + 1.21$

$x = 0\cdot67 \quad \text{and} \quad 1\cdot75$

Formula method for solving quadratics

The formula method is a development from completing the square. Let us start with the general form of the quadratic equation:

$$ax^2 + bx + c = 0$$

Now divide throughout by a to make the coefficient of x^2 unity, to give

$$x^2 + \frac{bx}{a} + \frac{c}{a} = 0$$

Transpose the constant to the right-hand side of the equation, i.e.

$$x^2 + \frac{bx}{a} = \frac{-c}{a}$$

Add the square of half the coefficient of x to both sides to complete the left-hand side:

$$x^2 + \frac{bx}{a} + \left(\frac{b}{2a}\right)^2 = \left(\frac{b}{2a}\right)^2 - \frac{c}{a}$$

This gives us

$$\left(x + \frac{b}{2a}\right)^2 = \frac{b^2 - 4ac}{4a^2}$$

If we now take the square root of both sides of the equation, we get

$$x + \frac{b}{2a} = \pm\frac{\sqrt{b^2 - 4ac}}{2a}$$

Now make x the subject of the equation:

$$x = \frac{-b \pm \sqrt{b^2 - 4ac}}{2a}$$

We can now say that the general formula for determining the roots of a quadratic equation is

$$x = \frac{-b \pm \sqrt{b^2 - 4ac}}{2a}$$

Using this formula we can get to the roots of an equation very quickly, but before we look at an example bear in mind the following points regarding its use:

1. All the terms must be gathered together and equated to zero $ax^2 + bx + c = 0$.

2. When we take the coefficient of x^2 we must include the sign.

3. When we take the coefficient of x we must include the sign.

4. When we take the constant we must include the sign.

EXAMPLE 1.31

Solve the following equation: $x^2 + x - 2 = 0$.

In this case, $a = 1$, $b = 1$ and $c = -2$.

$$\therefore \quad x = \frac{-1 \pm \sqrt{1^2 - 4(1)(-2)}}{2(1)}$$

$$x = \frac{-1 \pm 3}{2}$$

$$x = 1 \quad \text{or} \quad -2$$

EXAMPLE 1.32

Solve $4x^2 + 9x + 2 = 0$.

In this case, $a = 4$, $b = 9$ and $c = 2$.

$$\therefore \quad x = \frac{-9 \pm \sqrt{9^2 - 4(4)(2)}}{2(4)}$$

$$x = \frac{-9 \pm 7}{8}$$

$$x = -0.25 \quad \text{or} \quad -2.$$

EXAMPLE 1.33

Solve $x^2 - 10x + 24 = 0$.

$$x = \frac{-(-10) \pm \sqrt{(-10)^2 - 4(1)(24)}}{2(1)}$$

$$x = \frac{10 \pm 2}{2}$$

$$x = 4 \quad \text{or} \quad 6$$

EXAMPLE 1.34

Find the roots of $\dfrac{x^2}{6} - \dfrac{x}{3} - 4 = 0$.

$$\frac{x^2}{6} - \frac{x}{3} - 4 = 0$$

Multiply throughout by 6:

$$x^2 - 2x - 24 = 0$$

$$x = \frac{-(-2) \pm \sqrt{(-2)^2 - 4(1)(-24)}}{2(1)}$$

$$x = \frac{2 \pm 10}{2}$$

$$x = 6 \quad \text{or} \quad -4$$

If we now turn to our formula for the solution of quadratic equations, further examination will show us that there are three possible types of solution. Which of these types of solution will occur depends on the values of b and $4ac$.

SOLUTION TYPE 1

Solve $x^2 - 4x + 4 = 0$.

$$x = \frac{-(-4) \pm \sqrt{(-4)^2 - 4(1)(4)}}{2(1)}$$

$$x = \frac{4 \pm \sqrt{(16 - 16)}}{2}$$

$$x = 2$$

i.e. $b^2 = 4ac$.

SOLUTION TYPE 2

Solve $3x^2 - 5x - 2 = 0$.

$$x = \frac{-(-5) \pm \sqrt{(-5)^2 - 4(3)(-2)}}{2(3)}$$

$$x = \frac{5 \pm 7}{6}$$

$$x = 2 \quad \text{or} \quad -\frac{1}{3}$$

i.e. $b^2 > 4ac$.

SOLUTION TYPE 3

Solve $3x^2 + 2x + 4 = 0$.

$$x = \frac{-2 \pm \sqrt{2^2 - 4(3)(4)}}{2(3)}$$

$$x = \frac{-2 \pm \sqrt{-44}}{6}$$

Since we cannot find the root of -44 without the use of imaginary numbers, an area of mathematics that does not concern us, we can say that x has no real roots, i.e. $b^2 < 4ac$.

The quantity $b^2 - 4ac$ is called the discriminant of the equation and we can summarise the above solutions as follows:

- $b^2 - 4ac = 0$ has two equal roots
- $b^2 > 4ac$ has two distinct roots
- $b^2 < 4ac$ has no equal roots.

Quadratic problems

EXAMPLE 1.35

The span of a steel roof truss is $(2x + 14)$ m and the rise is x metres. The rafter length is given as $(2x + 1)$ m. Calculate the span of the roof and the rise of the truss.

Whenever possible it is always a good practice to sketch the problem (see Fig. 1.8).

Figure 1.8

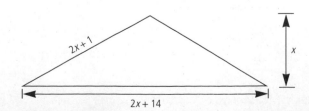

Considering half the roof truss we get the sketch shown in Fig. 1.9.

Figure 1.9

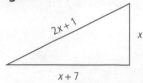

Applying the theorem of Pythagorus, we get

$$x^2 + (x+7)^2 = (2x+1)^2$$

$$x^2 + x^2 + 14x + 49 = 4x^2 + 4x + 1$$

$$2x^2 + 14x + 49 = 4x^2 + 4x + 1$$

$$-2x^2 + 10x + 48 = 0$$

$$x = \frac{-10 \pm \sqrt{10^2 - 4(-2)(48)}}{2(-2)}$$

$$x = \frac{-10 \pm 22}{-4} = 8 \quad \text{and} \quad -3$$

Since we cannot have a negative length then $x = 8$. Therefore the rise $= 8\,\text{m}$ and the span $2x + 14 = 30\,\text{m}$.

EXAMPLE 1.36

The surface area of a cylinder is given by $s = 22\pi r(h+r)$ where h is the height and r the radius. Given that the surface area $= 0.5\,\text{m}^2$ and the height is 300 mm, calculate the diameter of the cylinder.

The first point to note is that the dimensions are mixed, i.e. m^2 and mm. We must work with one or the other.

$$s = 0.5\,\text{m}^2 \quad h = 0.3\,\text{m}$$

$$s = 2\pi r(h+r)$$

$$s = 2\pi rh + 2\pi r^2$$

$$2\pi r^2 + 2\pi rh - s = 0$$

$$2(3.142)r^2 + 2(3.142)(0.3)r - 0.5 = 0$$

$$r = \frac{-1.885 \pm \sqrt{1.885^2 - 4(6.284)(-0.5)}}{2(6.284)}$$

$$r = \frac{-1.885 \pm 4.015}{12.568}$$

$$r = 0.169 \quad \text{and} \quad -0.469$$

Therefore the radius is $0.169\,\text{m}$ and the diameter $= 0.338\,\text{m}$ or 338 mm.

EXAMPLE 1.37

A concrete beam is reinforced with steel bars. The breadth of the beam is 300 mm and the depth d to the centre of the steel reinforcement is 800 mm. The cross-sectional area of the steel is 1000 mm². Calculate the depth of the neutral axis n from the top of the beam given the formula as

$$\frac{2a}{bn} = \frac{n}{d-n} \times \frac{1}{15}$$

Figure 1.10 shows a sketch of the problem

Figure 1.10

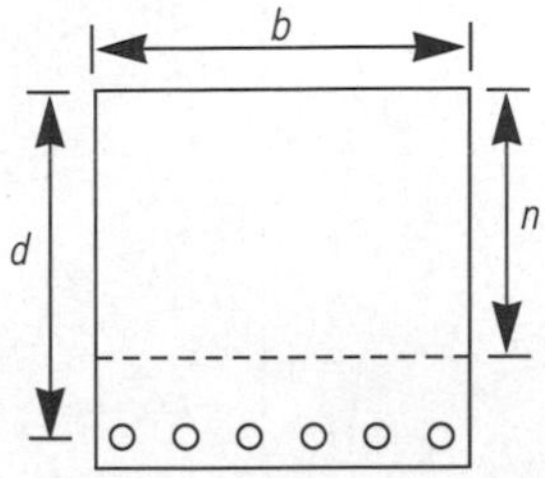

$a = 1000$
$b = 300$
$d = 800$

So, $\dfrac{2(1000)}{300n} = \dfrac{n}{800-n} \times \left(\dfrac{1}{15}\right)$

$$\frac{2000}{300n} = \frac{n}{12000 - 15n}$$

$$2000(12000 - 15n) = 300n^2$$

$$24(10^6) - 3(10^4)n = 3(10^2)n^2$$

$$8(10^4) - (10^2)n = n^2$$

$$n^2 + (10^2)n - 8(10^4) = 0$$

$$n^2 + 100n - 80\,000 = 0$$

Using the formula $\quad n = \dfrac{-b \pm \sqrt{b^2 - 4ac}}{2a}$

where $a = 1$, $b = 100$, $c = -80\,000$

we have $\quad n = \dfrac{-100 \pm \sqrt{(100)^2 - 4(1)(-80\,000)}}{2}$

$$= \frac{-100 \pm \sqrt{10^4 + 32 \times 10^4}}{2}$$

$$= \frac{-100 \pm 10^2\sqrt{1 + 32}}{2} = \frac{-100 \pm 10^2\sqrt{33}}{2}$$

$$= \frac{-100 \pm 574.4}{2}$$

∴ Taking positive value only $\quad n = \dfrac{474.4}{2} = 237.2\,\text{mm}$

1.4 Cubic equations

When an equation contains a term raised to the power of 3, it is said to be a third degree or more commonly a cubic equation. It takes the general form of $ax^3 + bx^2 + cx + d = y$, where a, b, c and d are constants. As with quadratic equations the cubic equation can be solved graphically. The graph of a cubic equation has two points where the curve turns. The curve of a cubic equation may cut the x axis at one, two or three points, which means there may be one, two or three roots to the particular equation. This is illustrated in Fig. 1.11.

Figure 1.11

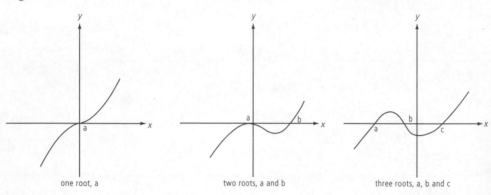

one root, a   two roots, a and b   three roots, a, b and c

When plotting the coordinates of a cubic equation it is a good idea to draw up a table similar to that shown when solving quadratics graphically.

EXAMPLE 1.38

Solve graphically $y = x^3 + 2x^2 - 3x - 4$ over the range -3 to 3.

The first step is to produce a table and calculate the value of x and y:

When	$x =$	-3	-2	-1	0	1	2	3
	$x^3 =$	-27	-8	-1	0	1	8	27
	$+2x^2 =$	18	8	2	0	2	8	18
	$-3x =$	9	6	3	0	-3	-6	-9
	$-4 =$	-4	-4	-4	-4	-4	-4	-4
then	$y =$	-4	2	0	-4	-4	6	32

We can now plot the graph (see Fig. 1.12). The roots of the equation are taken from where the curve crosses the x axis and these are approximately -2.6, -1 and 1.6.

Figure 1.12

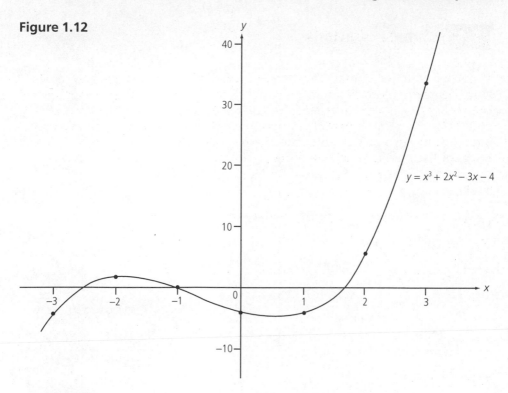

$$y = x^3 + 2x^2 - 3x - 4$$

EXAMPLE 1.39

Plot the graph of $x^3 - x = 0$ and find the solution to the equation.

When $x =$	-3	-2	-1	0	1	2	3
$x^3 =$	-27	-8	-1	0	1	8	27
$-x =$	3	2	1	0	-1	-2	-3
$y =$	-24	-6	0	0	0	6	24

From the graph shown in Fig. 1.13, we can see that the solutions are 0 and ± 1.

1.5 Progressions

A progression is a series of numbers which occur in various forms. Some of these forms are fairly simple and obvious, such as 2, 4, 6, 8, while others are not so obvious, such as 4, -6, 9, -13.5.

We can define progressions as a series of numbers in which any number has a constant relationship with the immediate preceding and succeeding numbers.

There is more than one type of progression, so we will consider each type in turn.

Figure 1.13

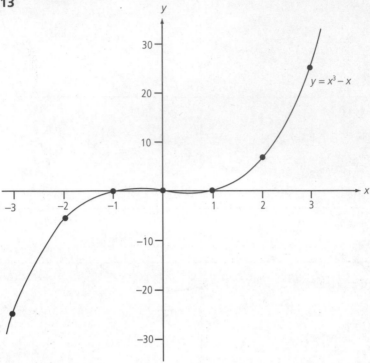

Arithmetic progression

This is probably the simplest of the progressions and is usually abbreviated to AP. In an arithmetic progression any term in the series can be found by adding a constant value to the preceding term, i.e. in the progression 2, 4, 6, 8,

$$2 + 2 = 4$$
$$4 + 2 = 6$$
$$6 + 2 = 8$$

In this case the constant value is 2.

We can now put this technique into a general form. By calling the first term a and the constant the common difference d the progression could be written in the form

a	$(a + d)$	$(a + 2d)$	$(a + 3d)$	$(a + (n - 1)d)$
1st	2nd	3rd	4th	nth term

If there are n terms in the series then the nth term is equal to $a + (n - 1)d$.

EXAMPLE 1.40

Write down the 10th term of the progression 2, 4, 6, . . .

First term $a = 2$

Common difference $d = 4 - 2 = 2$

$$\begin{aligned}
\text{10th term} &= a + (10 - 1)d \\
&= 2 + 9 \times 2 \\
&= 20
\end{aligned}$$

EXAMPLE 1.41

Find the common difference in the progression x, $4x$, $7x$ . . .

$$\begin{aligned}
\text{Common difference } d &= 4x - x \\
&= 3x
\end{aligned}$$

EXAMPLE 1.42

There are a total of five numbers in the progression starting at 3 and ending with 11. Find the three missing numbers.

$$\begin{aligned}
\text{First term} &= 3 \\
\text{Fifth term} &= a + 4d = 11 \\
&= 3 + 4d = 11
\end{aligned}$$

Therefore

$$\begin{aligned}
4d &= 11 - 3 = 8 \\
d &= 8 \div 4 = 2
\end{aligned}$$

EXAMPLE 1.43

An excavator is bought for £11 586 and depreciates at a rate of £540 per annum. What is its value at the end of 5 years?

At the end of the first year the value is £11 586 − 540 = £11 046 (1st term) and d is equal to −£540.

At the end of 5 years the value is £11 046 + (5 − 1) × (−540) = £8886.

Note! We are calculating values at the end of the year.

We are often required to find the summation of a progression. One way would be to carry out a long manual process, i.e. $a + (a + d) + (a + 2d)$ etc., which would be rather boring for, shall we say, 50 terms. An alternative method can be found from the following, where s = sum, a = the first term, d = the common difference, and n = the number of terms.

Equation (1) starts with the first term and increases to the last:

$$s = a + (a + d) + (a + 2d) + (a + 3d) + \ldots + (a + (n - 2)d) + (a + (n - 1)d) \quad (1)$$

Equation (2) starts with the last term and decreases to the first:

$$s + (a + (n - 1)d) + (a + (n - 2)d) + \ldots + (a + 3d) + (a + 2d) + (a + d) + a \quad (2)$$

Adding the two equations together:

$$2s = a + (2a + nd) + (2a + nd) + \ldots + (2a + nd) + (2a + nd) + a$$

$$2s = 2a + (2a + nd) + (2a + nd) + \ldots + (2a + nd) + (2a + nd)$$

$$2s = 2an + (n - 1)nd$$

$$2s = n(2a + (n - 1)d)$$

$$s = \frac{n}{2}(2a + (n - 1)d)$$

Therefore the summation of n terms is

$$s_n = \frac{n}{2}(2a + (n - 1)d)$$

EXAMPLE 1.44

Calculate the values of the 10th term and the sum to 19 terms of the progression
$1\cdot3 + 2\cdot8 + 4\cdot3 \ldots$.

$$a = 1\cdot3 \qquad d = 1\cdot5$$

$$\begin{aligned}
\text{10th term} &= a + (n - 1)d \\
&= 1\cdot3 + (10 - 1)\,1\cdot5 \\
&= 1\cdot3 + 9 \times 1\cdot5 \\
&= 1\cdot3 + 13\cdot5 \\
&= 14\cdot8
\end{aligned}$$

Summation of 19 terms:

$$s_{19} = \frac{n}{2}(2a + (n - 1)d)$$

$$s_{19} = \frac{19}{2}[2 \times 1\cdot3 + (19 - 1)\,1\cdot5]$$

$$s_{19} = 9\cdot5(2\cdot6 + 18 \times 1\cdot5)$$

$$s_{19} = 9\cdot5(2\cdot6 + 27)$$

$$s_{19} = 9\cdot5\,(29\cdot6)$$

$$s_{19} = 281\cdot2$$

EXAMPLE 1.45

A trainee starts a job at an annual salary of £4400 with an increase of £400 for each year of service.
How much will he have earned after 22 years' service?

$$s_{22} = \frac{n}{2}(2a + (n - 1)d)$$

where $n = 22$, $a = £4400$, $d = £400$.

$$S_{22} = \frac{22}{2}[2(4400) + (22 - 1)400]$$

$$S_{22} = 11(8800 + 8400)$$

$$S_{22} = 11(17\,200)$$

$$S_{22} = £189\,200$$

EXAMPLE 1.46

An estimate for a basement excavation 18 m × 10 m × 6 m deep has to be prepared for a proposed building. If the all-in rate for the first metre of depth is £24·50 per m³, and for each successive metre of depth increases by £2·50 per m³, calculate the cost of the excavation work.

METHOD 1

Taking an area of 1 m², then for a depth of 6 m, i.e. 6 m³,

$$s_n = \frac{n}{2}(2a + (n - 1)d)$$

$$s_6 = \frac{6}{2}(2(24·5) + (6 - 1)2·5)$$

$$s_6 = 3(49 + 12·5)$$

$$s_6 = 3(61·5)$$

$$s_6 = 184·5$$

Since the area is 18 × 10 = 180 m², then

$$\text{total cost} = 184·5\,(180)$$
$$= £33\,210$$

METHOD 2

$$\text{Area} = 18 \times 10 = 180\,\text{m}^2$$

$$d = 180(2.5) = 450$$

$$\text{Cost for 1 m depth} = 180\,(24·5) \quad = 4410$$

$$\text{Cost for 6 m depth } s_n = \frac{n}{2}\,(2a + (n - 1)d)$$

$$s_6 = \frac{6}{2}\,(2 \times 4410) + (6 - 1)450)$$

$$s_6 = 3(8820 + 2250)$$

$$s_6 = 3 \times 11\,070$$

$$s_6 = £33\,210$$

Simple interest

When monies are invested, the investor receives interest on these monies at a given rate. If this additional money is not reinvested then it is referred to as simple interest and the initial sum, called the capital or principal, does not increase.

If we call the money invested £P and the interest r%, since £r will be paid on every £100 invested, then after 1 year the interest paid will be

$$\frac{P}{100}(r) = \frac{Pr}{100}$$

This amount of interest will be paid each year until the invested money is withdrawn or the interest rates change. If we call the time invested t years, then the total interest paid will be

$$\frac{Pr}{100}(t) = \frac{Prt}{100}$$

The total value of the investment after a given number of years will be the amount of the original sum of money plus the interest, i.e.

$$£P + \frac{Prt}{100} = P\left(\frac{1 + rt}{100}\right)$$

EXAMPLE 1.47

How much simple interest would be paid to an investor after 5 years if the sum invested was £10 000 and the interest rate was 6%?

$$\text{Simple interest} = \frac{Prt}{100} \quad \text{where } P = 10\,000 \quad r = 6\% \quad t = 5$$

$$= \frac{10\,000 \times 6 \times 5}{100}$$

$$= £3000$$

EXAMPLE 1.48

A man invests £5600 at a simple interest rate of 7% for $9^1/_3$ years. Calculate the total value of his investment.

$$\text{Total value} = P\left(1 + \frac{rt}{100}\right)$$

$$= 5600\left(1 + \frac{7(28)}{100(3)}\right)$$

$$= 5600 + 3658.67$$

$$= £9258.67$$

EXAMPLE 1.49

A man receives the sum of £800 after 5 years as simple interest on an investment of £4500. What is the rate of interest?

If we call the simple interest I, then

$$I = \frac{Prt}{100}$$

$$Prt = 100I$$

$$r = \frac{100I}{Pt}$$

$$= \frac{100(800)}{4500(5)}$$

$$= 3.56\%$$

Geometric progressions (GPs)

A geometric progression is one in which the preceding number is multiplied by a constant, i.e. 2, 4, 8 . . . is a geometric progression in which 2 is the constant multiplier, i.e.

$$2(2) = 4$$

$$4(2) = 8$$

This constant is generally referred to as the common ratio r and may be either positive or negative. The ratio can be found by dividing any term by the preceding term, i.e.

$$\frac{4}{2} = 2$$

$$\frac{8}{4} = 2$$

To establish a formula for calculating our terms, let us call our first term a and the ratio r. We can then say

1st term $= a$

2nd term $= ar$

3rd term $= ar(r) = ar^2$ index of $r = 2$

4th term $= ar^2(r) = ar^3$ index of $r = 3$

5th term $= ar^3(r) = ar^4$ index of $r = 4$

nth term $= ar^{n-1}$ index of $r = n - 1$

We can say that the nth term of a progression is equal to ar^{n-1}, i.e. nth term $= ar^{n-1}$.

EXAMPLE 1.50

Find the eighth term of the progression 6, 18, 54 . . . .

$$1\text{st term} = 6$$

$$r = \frac{18}{6} = 3$$

$$n\text{th term} = ar^{n-1}$$

Therefore

$$8\text{th term} = 6(3^{8-1}) = 13\,122$$

EXAMPLE 1.51

Find the sixth term of 3, 6, 12 . . . .

$$a = 3$$

$$r = \frac{6}{3} = 2$$

$$6\text{th term} = 3(2^5) = 96$$

EXAMPLE 1.52

Find the fifth term of 9, 3, 1 . . . .

$$a = 9$$

$$r = \frac{3}{9} = \frac{1}{3}$$

$$5\text{th term} = 9\left(\frac{1}{3}\right)^4$$

$$= 0.11$$

EXAMPLE 1.53

Find the first term of a geometric progression to two places of decimals given that the third term is 35 and the seventh term is 4200.

$$3\text{rd term} = ar^2 = 35$$

$$7\text{th term} = ar^6 = 4200$$

Dividing the seventh term by the third, we get

$$\frac{ar^6}{ar^2} = \frac{4200}{35}$$

$$r^4 = \frac{4200}{35} = 120$$

$$r = \sqrt[4]{120} = 3{\cdot}31$$

$$\text{1st term} = \frac{ar^2}{r^2}$$

$$= \frac{35}{10.95} = 3.19$$

To find the sum of a geometrical progression we can say let s_n denote the sum to n terms. Then

$$s_n = a + ar + ar^2 + ar^3 + \ldots + ar^{n-1} \quad (1)$$

Now multiply throughout by r and we get

$$rs_n = ar + ar^2 + ar^3 + ar^4 + \ldots ar^n \quad (2)$$

If we now subtract (2) from (1), we find

$$s_n - rs_n = a - ar^n$$

Therefore

$$s_n(1 - r) = a(1 - r^n)$$

$$s_n = \frac{a(1 - r^n)}{1 - r}$$

Similarly if we had subtracted (1) from (2) we would have got

$$s_n = \frac{a(r^n - 1)}{r - 1}$$

Both formulae are correct, but

$$s_n = \frac{a(1 - r^n)}{1 - r}$$

is more convenient to use when r is less than 1 and

$$s_n = \frac{a(r^n - 1)}{r - 1}$$

is more convenient when r is greater than 1.

When r is less than 1 and n is large then r^n approaches zero, and if we say that n approaches infinity (∞) then r^n disappears.

Taking the formula

$$s_n = \frac{a(1-r^n)}{1-r}$$

we can say

$$s_n = \frac{a}{1-r} - \frac{ar^n}{1-r}$$

and as n approaches infinity the term ar disappears, leaving us with

$$s_n = \frac{a}{1-r}$$

This is called the sum to infinity of a geometrical progression.

EXAMPLE 1.54

Find the sum to eight terms of the series 6, 9, 13.5 . . . .

$$a = 6 \qquad r = \frac{9}{6} = 1.5$$

Since r is greater than 1, we use

$$s_8 = \frac{a(r^n - 1)}{r - 1}$$

$$= \frac{6(1.5^8 - 1)}{1.5 - 1}$$

$$= 295.55$$

EXAMPLE 1.55

Find the sum to infinity of the series 2, 1, 1/2 . . . .

$$a = 2 \qquad r = \frac{1}{2} = 0.5$$

$$s_\infty = \frac{a}{1-r}$$

$$= \frac{2}{1 - 0.5}$$

$$= 4$$

EXAMPLE 1.56

Find the 10th term and the sum of the first eight terms of 1.7, 3.6, 4.8 . . . .

$$a = 2.7 \qquad r = \frac{3.6}{2.7} = 1.33$$

$$S_8 = \frac{a\,(r^n - 1)}{r - 1}$$

$$= \frac{3.6(1.33^8 - 1)}{1.33 - 1}$$

$$= 95.9$$

$$\text{10th term} = ar^{10-1}$$

$$= 2.7(1.33^{10-1})$$

$$= 35.16$$

EXAMPLE 1.57

Find the first term and the sum of the first five terms of the GP whose third term is 27 and whose fifth term is 3.

$$ar^2 = 27 \qquad ar^4 = 3$$

$$\frac{ar^2}{ar^4} = \frac{27}{3}$$

$$\frac{1}{r^2} = \frac{27}{3}$$

$$r = \sqrt{3/27} = \frac{1}{3}$$

$$a = \frac{ar^2}{r^2} = 27 \div \frac{1}{9} = 243$$

Sum of the five terms:

$$S_5 = \frac{a\,(1 - r^n)}{1 - r}$$

$$= \frac{243(1 - 0.33^5)}{1 - 0.33}$$

$$= 360.27$$

EXAMPLE 1.58

A pile boring machine bores 4 m in the first hour and then the rate of boring decreases by 15% an hour. Find

(a) the depth of the bore hole after 3 hours
(b) the time taken to bore a depth of 16 m.

The rate of boring forms a GP where $a = 4$ m and $r = 85\%$ or 0.85:

$$S_n = \frac{a\,(1 - r^n)}{1 - r}$$

$$S_3 = \frac{4(1 - 0.85^3)}{1 - 0.85}$$

$$S_3 = \frac{4(1 - 0.614\,125)}{0.15}$$

$$S_3 = \frac{1.5435}{0.15}$$

$$S_3 = 10.29 \text{ m bored in 3 hours}$$

The number of hours taken to bore 16 m is n. Therefore

$$S = \frac{a(1 - r^n)}{1 - r}$$

$$16 = \frac{4(1 - 0.85^n)}{1 - 0.85}$$

$$16 = \frac{4(1 - 0.85^n)}{0.15}$$

$$2.4 = 4(1 - 0.85^n)$$

$$0.6 = 1 - 0.85^n$$

$$-0.4 = -0.85^n$$

$$0.85^n = 0.4$$

Taking logarithms of both sides,

$$\log(0.85^n) = \log 0.4$$

$$n \log 0.85 = \log 0.4$$

$$n = \frac{\log 0.4}{\log 0.85}$$

$$n = \frac{-0.3979}{-0.07\,058}$$

$$n = 5.638 \text{ hours}$$

We could extend this problem by asking: what is the maximum depth the machine can bore?

Maximum depth is equal to the sum to infinity:

$$S_\infty = \frac{a}{1 - r}$$

$$S_\infty = \frac{4}{1 - 0.85}$$

$$S_\infty = \frac{4}{0.15}$$

$$S_\infty = 26.67 \text{ m}$$

EXAMPLE 1.59

The preventative maintenance costs of a bungalow are estimated as being £250 in the first year with an annual increase of 7.5% for the following 25 years. Calculate the projected costs for the 15th year and also the total costs for the 25 years.

Projected cost for year 15 is ar^{n-1}, where $a = 250$ and $r = 107.5\%$ or 1.075.

$$\text{15th year costs} = 250(1.075^{14})$$

$$= 688.11$$

$$\text{Total costs} = \frac{a(r^n - 1)}{r - 1}$$

$$= \frac{250(1.075^{25} - 1)}{1.075 - 1}$$

$$= \frac{250(5.098)}{0.075}$$

$$= 16\,994.47$$

Compound interest and mortgages

If a sum of money is invested in say a building society account then an annual rate of interest will be added to the sum.

For example, let us say we invest £1000 at an annual rate of interest of 10%. Then at the end of the first year we get 10% of £1000 added to our investment to give a total of £1100. At the end of the second year we get 10% of £1100 added to our investment to give a total of £1210.

This will continue for as long as the money remains in the account. This then is compound interest and a simple definition would be to say we get interest on the interest.

Note: This is a simple example to show the basic method of compound interest. In practice the rates of compound interest will vary according to the type of investment and may also be paid on variable time periods, i.e. monthly, half-yearly.

If we call the sum invested the principal P and the interest rate r, then we can say after year 1 that the investment will be

$$£\left(P + \frac{Pr}{100}\right)$$

or $£P(1 + r/100)$, the new principal. After the second year the investment will be

$$P(1 + r/100) + [P(1 + r/100)(r/100)]$$

$$= P(1 + r/100)(1 + r/100)$$

$$= P(1 + r/100)^2$$

After 3 years the amount (A) of the investment will be $P(1 + r/100)^3$.

After n years the amount (A) of the investment will be $P(1 + r/100)^n$.

We can now say that at the end of a period of time the investment amount will be equal to

$$A = P(1 + r/100)^n$$

EXAMPLE 1.60

What will be the amount of an investment of £5000 invested at a 5% compound interest after a period of 10 years?

$$A = P(1 + r/100)^n$$

where $P = 5000$, $r = 5$, $n = 10$.

$$A = 5000(1 + 5/100)^{10}$$

$$A = 5000(1 + 0.05)^{10}$$

$$A = 8144.47$$

As you can see these calculations are very similar to geometrical progressions.

EXAMPLE 1.61

How long will it take for a sum of money to treble its value at 7.5% per annum compound interest?

The final amount will be equal to $3P$.

$$A = P(1 + r/100)^n$$

$$3P = P(1 + r/100)^n$$

$$3P = P(1.075)^n$$

$$3 = 1.075^n$$

$$\log 3 = n \log 1.075$$

$$n = \log 3/\log 1.075$$

$$= 15.19 \text{ years}$$

The same technique can be applied to depreciation using the formula $A = P(1 - r/100)^n$. With this formula we are subtracting the depreciation rate from the annual amount.

EXAMPLE 1.62

A company car costing £12 000 when new depreciates at the rate of 20% per year. What will be its value after 3 years?

$$\text{Value} = P(1 - r/100)^n$$

$$= 12\,000(1 - 20/100)^3$$

$$= 12\,000(0.8)^3$$

∴ Value = £6144

Mortgage repayments

For the majority of people probably the largest debt they incur is that of a mortgage to buy a domestic dwelling. This sum of money is paid back over a number of years, with the total amount to repay calculated at compound interest. If the capital borrowed is £C and the interest rate is r% for a period of time t, then the total value of the money to be repaid will be

$$C(1 + r/100)^t$$

If the amount (A) is to be paid in arrears then the total amount will be the geometrical series

$$\text{total} = \frac{A[(1 + (r/100)^t] - 1)}{(1 + r/100)}$$

We can see where this formula comes from by comparing it to

$$S_n = \frac{a(r^n - 1)}{r - 1}$$

i.e. total amount $= s$

$$\text{payment } A = a$$

$$1 + r/100 = \text{common ratio } r$$

$$(1 + r/100)^t = r$$

These two equations are equal since both are the total repayments; hence

$$C(1 + r/100)^t = \frac{A[(1 + (r/100)]^t - 1)}{(1 + r/100) - 1}$$

Therefore capital borrowed

$$C = A\left(1 - \frac{1}{(1 + r/100)^t}\middle/ (r/100)\right)$$

and the repayment

$$A = \frac{C[1 + (r/100)]^t(r/100)}{[(1 + (r/100)]^t - 1)}$$

EXAMPLE 1.63

The purchase price for a new bungalow is £69 000. The purchaser agrees to pay a deposit of £9000 and borrow the remainder from a building society over a 25 year period. If the interest rate is 7%, calculate the purchaser's monthly repayments.

Amount borrowed C = £69 000 – 9000

$$= £60\ 000$$

Annual repayments $A = \dfrac{C(1 + r/100)^t(r/100)}{[1 + (r/100)]^t - 1}$

where $t = 25$ years and $r = 7\%$.

$$A = \dfrac{60\ 000(1.07)^{25}(0.07)}{(1.07)^{25} - 1}$$

$$A = 5148.63$$

Monthly repayments $= 5148.63/12$

$$= £429.05$$

EXAMPLE 1.64

The compound interest rate for a mortgage is 8% and a prospective borrower decides that he can afford repayments of £2400 per annum over a 20 year period. What is the amount he can borrow?

Capital borrowed $C = A\left(\dfrac{1 - \dfrac{1}{(1 + r/100)^t}}{r/100}\right)$

where $A = 2400$, $r = 8\%$, $t = 20$ years

then $C = 2400\left(\dfrac{1 - \dfrac{1}{(1.08)^{20}}}{0.08}\right)$

$$= 30\ 000\left(1 - \dfrac{1}{(1.08)^{20}}\right)$$

$$= £23\ 563.55$$

1.6 Application exercises

Linear equations

Solve the following equations:

1 $3(x + 18) = 18$

2 $9(3y - 8) + 7(6y + 4) = 2(8y + 7) - 3(4y - 9)$

3 $3(12x + 8) = 4(12 - 3x)$

4 $0.25m + m = 25$

5 $0.2x + 0.7x = 27$

6 A businessman spent a total of £182 500 on buying a building and

converting it into a factory. If the conversion comprised 0.125 of the buying price and the equipment cost 0.4 of the buying price, what did the businessman pay for the property?

7 Eight skilled men and three unskilled men were employed on a building site. If the skilled men were paid £0.75 more per hour than the unskilled men, what was the rate of pay for a skilled man if the total wage bill for 1 hour was £61.00?

8 In an isosceles triangle the equal angles are 2.5 times the size of the third angle. Calculate the size of this angle.

9 The perimeter of a plot of land is 9600 m and the length is five times the width. Find the area of the plot in hectares.

10 A liquid solution contains 5% water by volume. It is required to be diluted further until it contains 20% water by volume. How many litres of water must be added to 80 litres of the original mixture?

Simultaneous linear equations

Solve by graphical methods:

1 $2a + 3b = 5$
  $a + b = 2$

2 $4a - 4b = 21$
  $4a + 4b = 27$

3 $4m + n = 27$
  $2m + n = 17$

4 $3f - g = 15$
  $f + g = 17$

Solve by substitution:

1 $4x + y = 27$
  $2x + y = 15$

2 $3x - y = 15$
  $x + y = 17$

3 $2a + 3b = 40$
  $2a - 2b = 10$

4 $3r - 6s = 12$
  $3r + 2s = 36$

5 $4m - 3n = 15$
  $4m - 5n = 9$

6 $7x + 4y = 79$
  $4x + 3y = 48$

7 $4f + 5g = 13$
  $5f + 7g = 14$

8 $9x + 7y = -3$
  $5x - 3y = 19$

9 $11m - 7y = 55$
  $13m + 6y = 222$

10 $10x = 7x + 4y = -40$

Solve by elimination:

1 $2x + 3y = 5.2$
  $3x - 2y = 0$

2 $2a + b = 7$
  $a - 5b = 9$

3 $4p + 5q = 42$
  $10p + 3q = 20$

4 $6x - 21y = 7$
  $19x - 3y = 4$

5 $1.2x + y = 1.8$
  $x - 1.2y = 3.94$

6 $\quad p + 2q = -2$
  $1.5p - 0.4q = 10.6$

7 $\dfrac{x}{2} + \dfrac{y}{3} = 7$

  $\dfrac{x}{3} - \dfrac{y}{4} = -1$

8 $\dfrac{a}{5} + \dfrac{b}{3} = 4$

$\dfrac{a}{2} + \dfrac{b}{4} = 3.5$

9 $4m = \dfrac{n-3}{5} + 8$

$5n = 15 - \dfrac{m-2}{3}$

10 $1 - (4a + 5b) = 7a + 9b - 1 = 0$

11 $\dfrac{1}{x} - \dfrac{2}{y} = 12$

$\dfrac{1}{x} - \dfrac{3}{y} = 2$

12 $\dfrac{3}{r} - \dfrac{4}{s} = 5$

$\dfrac{1}{r} + \dfrac{3}{s} = 1$

Solve the following problems by simultaneous equations:

1 A builder receives two invoices for goods purchased. One invoice shows he bought five boxes of 50×8 screws and eight boxes of 40×6 screws for a total cost of £50.58. The second invoice shows he bought seven boxes of 50×8 screws and four boxes of 40×6 screws for £44.60. Find the cost per box of each size of screw.

2 If 2 tonne of cement and 0.5 tonne of lime cost £229.80, and 6 tonne of cement and 1 tonne of lime cost £655.20, find the cost of a 50 kg of cement and a 50 kg bag of lime.

3 The total cross-sectional area of one small pipe and two larger pipes is 5061 mm², and the total cross-sectional area of two small pipes and one large pipe is 4231.5 mm². Find the diameters of each size of pipe.

4 An order for 1500 facing bricks and 160 insulation blocks costs £407. Another order for 750 facing bricks and 100 insulation blocks costs £207.50. Calculate the cost of the facing bricks per 1000 and the cost of insulation blocks per m², given that there are 10 blocks/m².

5 A flight of stairs with a total rise of $0.75x$ and a total going of $0.6y$ requires a 4.2 m length of carpet to cover it. A second flight with a total rise of $0.875x$ and a total going of $0.5y$ requires a 4.24 m length of carpet to cover it. Calculate the rise and going of each flight.

Quadratic equations

Solve graphically the following equations to one place of decimals:

1 $3x^2 - 5x - 4 = 0$

2 $4x^2 + 3x - 5 = 0$

3 $y = x^2 - 5x + 2.25$

4 $y = 5x - 3x^2 + 7$

5 $y = x^2 - 8x + 7$

6 $y = x(x + 7) + 7.4$

7 Plot the graph of $5x^2 + 2x - 6 = 0$ and use this graph to solve $5x^2 + 2x - 8 = 0$.

8 Solve the simultaneous equations $y = x^2 - 4$ and $y = 2x + 1$.

9 Solve the simultaneous equations $y = 3x^2 + 3.5x - 7.25$ and $y = -0.5x - 2.25$.

Solve by factorisation:

1 $x^2 + 3x + 2 = 0$

2 $x^2 - 5x + 6 = 0$

3 $3x^2 + x - 2 = 0$

4 $x^2 - x - 2 = 0$

5 $x^2 + 2x - 3 = 0$

6 $4x - 1 = 4x^2$

7 $6x^2 - 11x + 3 = 0$

8 $10x^2 - 9 = 13x$

9 $14x^2 + 12 = 29x$

10 $x^2 - 12x = 0$

Solve by completing the square correct to two decimal places:

1 $x^2 - 17x + 66 = 0$

2 $x^2 + x - 6 = 0$

3 $2x^2 + 7x - 4 = 0$

4 $6x^2 + 4 = 11x$

5 $6x^2 + 5x - 6 = 0$

6 $2x^2 = 9x - 5$

7 $15x^2 = 11x + 12$

8 $4x^2 = 11x + 34$

9 $x + 5 = 6(x^2 - 5)$

10 $(x + 1)(5x + 12) = 24$

Solve using the formula method correct to two decimal places:

1 $2x^2 + 3x = 20$

2 $x^2 + x - 2 = 0$

3 $3x^2 + 4x - 1 = 0$

4 $x^2 + 7x + 12 = 0$

5 $5x^2 + 24x = 5$

6 $-5x = -x^2 - 3$

7 $(x + 4)(x - 5) = 18x$

8 $x(x + 7) = 7(x + 28)$

9 $x(9x - 2) = 2(x + 8)$

10 $\dfrac{1}{x - 1} + \dfrac{1}{x + 1} = 2$

Solve the following problems:

1 A freely supported beam carries a uniformly distributed load. The following formula will give the bending moment at a point along the beam. Given that w, the weight/metre, is 100 kg, l the length of the beam is 5 m, B the bending moment is 75 kg m and the formula is $B = wx(l - x)/2$, find x, the distance of the point from a support.

2 A pipe has a diameter of $(2x - 1)$ mm and will discharge the same volume of water as two other pipes with diameters of $(x + 5)$ mm and $(x - 2)$ mm. Find the diameters of the three pipes.

3 Given that $0.4t = 4c + c^2$, find the value of c when $t = 7.5$.

4 A can of paint will cover a square area whose length of side is x metres. The same volume of paint will also cover a border 0.75 m wide to a square area whose length of side is $x + 6$ metres. Find the length of the original square.

5 A brick wall is seven times longer than its height. If the height is increased by 1m and the length by 12 m, the area of the wall would be doubled. Calculate the number of bricks required to build the wall if there are 120 bricks per square metre.

Cubic equations

Solve graphically the following cubic equations:

1 $x^3 - 2x^2 - x + 2 = 0$ **4** $x^3 - 3x^2 - x + 3 = 0$

2 $x^3 - 3x^2 - 18x + 40 = 0$ **5** $x^3 - 4x^2 = 3x - 8$

3 $x^3 + x^2 + 2 = 5x$ **6** $x^3 - x^2 - 5x + 2 = 0$

Progressions

ARITHMETIC PROGRESSIONS

1 Find the 12th term of 65, 62, 59 . . . .

2 Find the eighth term of 24.5, 23, 21.5 . . . .

3 Find the 40th term of –29, –23, –17 . . . .

4 A company car is bought for £12 000 and depreciates at a rate of £1100 per annum. Calculate its value at the end of 4 years.

5 A series of bored piles 24 m long have to be bored through stiff clay subsoil. The cost of boring a pile is determined as follows. The first 3 m costs £22 and each successive 3 m increases in cost by £6. Determine:

(a) the cost of one pile
(b) the cost of 20 piles, allowing 20% for profits and overheads.

6 A construction technician starts work in a contractor's office earning £5000 per annum with an annual increase of £250. Four years later, after successfully completing the Higher National Certificate, he receives an increase of £500 on his salary with an annual increase of £350. How much does he earn after 10 years' service?

7 A local authority receives five deposits of £5000 each for tender documents which it invests for a period of 3 months at an annual rate of 5%. How much money does the local authority gain from this simple interest investment?

8 A new hospital costs £3 000 000 to build on a flexible price contract. If the cost of labour, plant and materials increased by 6.5% per annum over the contract period of 4 years, what would be the final cost of the hospital?

GEOMETRIC PROGESSIONS

1 Find the fifth term and the sum of the first eight terms of the following geometric series:

(a) 3.0   4.5   6.75

(b) 2   4   8

(c) 2.0   1.5   1.125

(d) 2.7   3.6   4.8

(e) 125   25   5

2 Sum to infinity:

(a) 2   1   0.5

(b) 30   10   3.33

(c) 48   24   12

(d) 2   0.5   0.125

(e) 5   −1   0.2

3 The annual output from a quarry producing stone ballast is 25 000 tonnes, giving a total income of £375 000. It has been decided that when the income is less than £200 000 the working of the quarry will no longer be viable. Given that output is decreasing by 10% per year, how much longer will the quarry continue to operate?

4 A site manager earning £12 000 a year estimates that his salary will increase by 5% annually. How much will he be earning in 7 years' time?

5 A rubbish chute attached along the side of a scaffold has a maximum height of 18 metres. Its lower end where it empties directly into the skip is 2 metres high. So that the chute will not block but will discharge slowly into the lorry, to avoid excessive dust it is supported by a series of eight props forming a geometrical progession. Given that the props cost £4.50 per metre, calculate the total cost of the props.

COMPOUND INTEREST AND MORTGAGE REPAYMENTS

1 A contractor decides that in 5 years' time he will need to replace some large items of plant. In order to meet these replacement costs, he invests £12 000 at a compound interest rate of 8%. What will be the final value of his investment in 5 years?

2 A company's excavating plant depreciates at the rate of 15% per annum. If the total value is £120 000 what will be its value in 3 years' time?

3 A new bungalow is going to cost the prospective buyer £89 000. The buyer estimates that he can afford a 15% deposit. A building society offers to lend him the rest of the money at 8% compound interest for a period of 15 years. How much will his annual mortgage be?

4 A company decides to invest 5% of its present profits into the stock market each year. The company profits continue to increase by 12.5% annually and after 5 years the board decides to change its investment to 5% of the present profits. If the annual return on the investment is 16% compound interest, calculate the total value of the investment after 10 years if in the initial year the company's profits were £500 000.

5 A contractor needs to increase his vehicle fleet by three new vehicles. The total cost is £58 000 and the deal he has been offered is 1/3 deposit and the remainder at 15% simple interest over 3 years, or 1/4 deposit and pay the remainder at 9% compound interest over 3 years. Which option will be the cheapest and by how much?

1.7 Answers to Section 1.6

Linear equations

1	$x = -12$	6	£119 672.13
2	$y = 1.308$	7	£5.75
3	$x = 0.5$	8	30°
4	$m = 20$	9	320 hectares
5	$x = 30$	10	15 litres

Simultaneous linear equations

GRAPHICAL SOLUTIONS

1	$a = 1$	$b = 1$	3	$m = 5$	$n = 7$
2	$a = 6$	$b = 0.75$	4	$f = 8$	$g = 9$

SUBSTITUTION SOLUTIONS

1	$x = 6$	$y = 3$	6	$x = 9$	$y = 4$
2	$x = 8$	$y = 9$	7	$f = 7$	$g = -3$
3	$a = 11$	$b = 6$	8	$x = 2$	$y = -3$
4	$r = 10$	$s = 3$	9	$m = 12$	$y = 11$
5	$m = 6$	$n = 3$	10	$x = -4$	$y = -3$

ELIMINATION SOLUTIONS

1	$x = 0.8$	$y = 1.2$	7	$x = 6$	$y = 12$
2	$a = 4$	$b = -1$	8	$a = 1.42$	$b = 11.14$
3	$p = -0.68$	$q = 8.94$	9	$m = 2$	$n = 3$
4	$x = 0.17$	$y = -0.29$	10	$a = 4$	$b = -3$
5	$x = 2.5$	$y = -1.2$	11	$n = 1/32$	$y = 1/10$
6	$p = 6$	$q = -4$	12	$r = 0.68$	$s = -6.5$

LINEAR PROBLEMS

1	£4.29	£3.64	4	£250	£2.00
2	£4.89	£3.42	5	2.09 m	2.15 m
3	50 mm	38 mm		2.44 m	1.795 m

Quadratic equations

GRAPHIC SOLUTIONS

1	2.3	−0.6	**6**	−5.7	−1.3
2	0.8	−1.6	**7**	−1.5	1.1
3	0.5	4.5	**8**	3.4	−1.4
4	−0.9	2.6	**9**	0.8	−2.1
5	7	1			

FACTORISATION SOLUTIONS

1	−1	−2	**6**	0.5	0.5
2	3	2	**7**	1/3	3/2
3	0.67	−1	**8**	9/5	−1/2
4	2	−1	**9**	3/2	4/7
5	−3	1	**10**	0	12

COMPLETING THE SQUARE SOLUTIONS

1	6	11	**6**	3.85	0.65
2	2	−3	**7**	1.3	−0.6
3	0.5	−4	**8**	4.6	−1.85
4	1.33	0.5	**9**	2.5	−2.33
5	0.67	−1.5	**10**	0.6	−4

FORMULA SOLUTIONS

1	2.5	−4	**6**	4.3	0.7
2	1	−2	**7**	20	−1
3	0.22	−1.55	**8**	14	−14
4	−3	−4	**9**	1.57	−1.13
5	0.2	−5	**10**	1.62	−0.62

SOLUTIONS TO PROBLEMS

1	0.321 mm	4.679 mm	**4**	6.24
2	5, 12 and 13 mm		**5**	8834 bricks
3	−4.65	0.65		

Cubic equations

GRAPHICAL SOLUTIONS

1	±1	2		**4**	±1	3	
2	2	5	−4	**5**	−1.65	1.2	4.3
3	−2.9	0.5	−1.4	**6**	2.6	0.38	−2.0

Progressions

ARITHMETIC PROGRESSIONS

1 32

2 14

3 205

4 £7600

5 (a) £344   (b) £7740

6 £8000

7 £312.50

8 £3 780 000

GEOMETRIC PROGRESSIONS

1 (a) 15.188   147.77
   (b) 32   510
   (c) 0.633   7.199
   (d) 8.533   72.808
   (e) 0.2   156.25

2 (a) 4
   (b) 45
   (c) 96
   (d) 2.67
   (e) 4.167

3 5.97 years

4 £16 885.21

5 £276.26

COMPOUND INTEREST AND MORTGAGE REPAYMENTS

1 £17 631.94

2 £73 695

3 £8838.16

4 £147 130.64

5 Compound interest option by £13.24

chapter

2

Mensuration techniques

Outcomes

At the end of this chapter you should be able to:

- calculate regular and irregular areas
- determine costs using simple interest techniques and volumes using various techniques including
  - trapezoidal rule
  - mid-ordinate rule
  - Simpson's rule
- apply the theorem of Pappus to solving rotating areas and volumes.

2.1   Areas and volumes

The methods used for finding the areas of regular figures such as rectangles, triangles, parallelograms and circles should already be familiar to us.

To briefly recap on these formulae:

- area of a rectangle = length × breadth

- area of a triangle = $\frac{1}{2}$ × base × perpendicular height

- area of a parallelogram = base × perpendicular height

- area of a circle = π × the radius squared.

We can now look at two further formulae for finding the area of a triangle.

1. To find the area of a triangle given two sides and the included angle as shown in Fig. 2.1, where a and b are known and θ (theta) is the included angle. We use the following formula:

$$\text{area } (A) = \frac{1}{2} ab \sin C$$

Figure 2.1

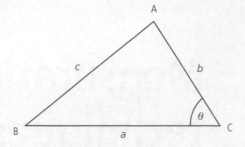

EXAMPLE 2.1

Given that two sides of a triangle are 80 mm and 102 mm and the included angle is 74°, calculate the area.

$$\text{Area} = \frac{1}{2} ab \sin C$$

where $a = 80$ mm, $b = 102$ mm and angle $C = 74°$

$$\text{Area} = 0.5 \times 80 \times 102 \times \sin 74$$

$$= 3921.95 \text{ mm}^2$$

---

2. To find the area of a triangle given the three sides as shown in Fig.2.2 where a, b and c are known, we use the following formula:

$$\text{area} = \sqrt{(s)(s-a)(s-b)(s-c)}$$

where a, b and c are the three sides and s = half the sum of the three sides.

Figure 2.2

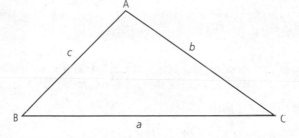

EXAMPLE 2.2

Find the area of the triangle whose sides are 50 mm, 65 mm and 95 mm.

$$\text{Area} = \sqrt{(s)(s-a)(s-b)(s-c)}$$

where $a = 50$ mm, $b = 65$ mm, $c = 95$ mm and

$$s = \frac{50 + 65 + 95}{2} = \frac{210}{2} = 105 \text{ mm}$$

$$\text{Area} = \sqrt{(105)(105-50)(105-65)(105-95)}$$

$$= 2\,310\,000 \text{ mm}^2$$

See Section 2.2 for further practical applications.

Areas of curved surfaces

THE CYLINDER

If we think of a cylinder as a pipe with very thin walls and we cut this pipe along its length and roll it out then we will get a rectangular shape. This is known as the development of the cylinder (see Fig. 2.3). The size of the rectangular shape is $2\pi r$ by l or h, where l is the length of the pipe or h is the height of the pipe.

Figure 2.3

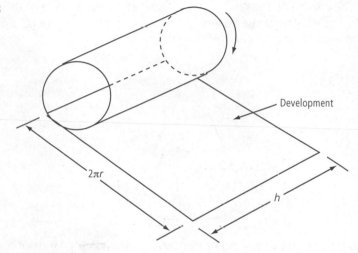

Therefore the surface area of the pipe is equal to $2\pi rh$.

If you now think of the pipe as being a solid cylinder then as well as the surface area around the cylinder we also have the surface area of the ends.

The area of one of the ends is equal to πr^2, so we can say that the total surface area of a solid cylinder is equal to the curved area + the area of the two ends, i.e.

surface area $= 2\pi rh + 2\pi r^2$

$\qquad\quad = 2\pi r\,(h + r)$

EXAMPLE 2.3

Calculate the surface area of a pipe of diameter 68 mm, and length 1.5 m.

Surface area $= 2\pi rh$ where $r = 68/2 = 34$ mm and $h = 1500$ mm

$\qquad\qquad\quad = 2\pi \times 34 \times 1500$

$\qquad\qquad\quad = 320\,442.45$ mm^2

EXAMPLE 2.4

Find the total surface area of a solid cylindrical bar of mild steel whose radius is 50 mm and length is 750 mm.

Total surface area $= 2\pi r(h+r)$

where $h = 750\,\text{mm}$ and $r = 50\,\text{mm}$.

Total surface area $= 2\pi 50(750 + 50)$

$$= 235\,619.45 + 15\,707.96$$

$$= 251\,327.41\,\text{mm}^2$$

THE CONE

This is a pyramid shape that has a circular base; see Fig. 2.4 where O represents the centre of the circular base, h represents the perpendicular height from the centre of the base and l represents the slant height of the cone from the edge of the base to the apex.

Figure 2.4

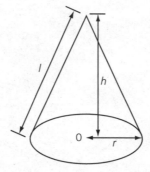

If we develop the cone in a similar manner to the development of a cylinder then we get a sector of a circle whose radius is equal to l and where length of arc is equal to $2\pi r$, the circumference of the base. This sector represents (see Fig. 2.5) the surface area of the cone.

Figure 2.5

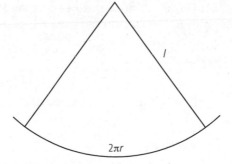

The area of a sector of a circle is given by

$$\frac{\text{arc length of the sector}}{\text{circumference of the circle}} \times \text{area of the circle}$$

$$= \frac{2\pi r}{2\pi l} \times \pi l^2$$

$$= \pi r l$$

So we can say that the curved surface of a cone is equal to half the circumference of the base multiplied by the slant height.

EXAMPLE 2.5

Calculate the curved surface area of a cone whose base diameter is 100 mm and whose slant height is 200 mm.

$$\text{Surface area} = \pi r l$$

where $r = 100/2 = 50$ mm and $l = 200$ mm.

$$\text{Surface area} = \pi \times 50 \times 200$$
$$= 31\,415.93 \text{ mm}^2$$

If the vertical height h and the radius of the base r are known, then we need to calculate the slant height l by using the theorem of Pythagoras. Then

$$l^2 = h^2 + r^2$$

Therefore

$$l = \sqrt{h^2 + r^2}$$

So that

$$\pi r l = \pi r \sqrt{h^2 + r^2}$$

The total surface area of a cone is equal to the curved surface area of the cone plus the area of the base, i.e.

$$\pi r l + \pi r^2 = \pi r (l + r)$$

EXAMPLE 2.6

Calculate the curved surface area of a cone whose base radius is 40 mm and perpendicular height is 80 mm.

METHOD 1

$$l = \sqrt{h^2 + r^2}$$
$$= \sqrt{80^2 + 40^2}$$
$$= 89.44 \text{ mm}$$

$$\text{Curved surface area} = \pi r l$$
$$= \pi \times 40 \times 89$$
$$= 11\,239.36 \text{ mm}^2$$

METHOD 2

$$\text{Curved surface area} = \pi r \sqrt{h^2 + r^2}$$
$$= \pi \times 40 \times \sqrt{80^2 + 40^2}$$
$$= 11\,239.70 \text{ mm}^2$$

The slight difference in the decimal figures is due to rounding errors.

EXAMPLE 2.7

Calculate the total surface area of a cone of perpendicular height 120 mm and diameter of the base 100 mm.

Total surface area $= \pi r(l + r)$

where $r = 100/2 = 50$ mm, and $l = \sqrt{h^2 + r^2}$, where $h = 120$ mm, i.e. $l = \sqrt{120^2 + 50^2} = 130$ mm.

$$
\begin{aligned}
\text{Total surface area} &= \pi r(l + r) \\
&= \pi \times 50 (130 + 50) \\
&= 20\,420.35 + 7853.98 \\
&= 28\,274.33 \text{ mm}^2
\end{aligned}
$$

See Section 2.2 for further practical applications.

The pyramid

This is a solid with a plane figure for its end whose faces form triangles that meet at a point. The base can be a polygon of any shape, including a circle, in which case we call it a cone as we already know. If all the edges joining the apex to the base are the same length then we call it a right pyramid. The length joining the apex to the centre of one base side is called the slant height (see Fig. 2.6).

Figure 2.6

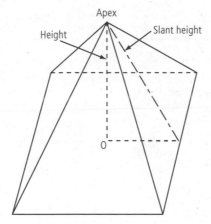

The surface area of a pyramid is the sum of the areas of the triangles forming the sides of the pyramid. The total surface area of a pyramid also includes the base.

EXAMPLE 2.8

Find the surface area of a pyramid having a height of 8 m and a square base of side 9 m (see Fig. 2.7).

Using the theorem of Pythagoras to find the slant height:

Figure 2.7

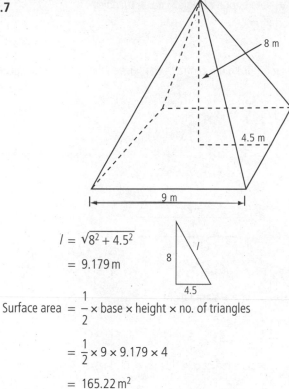

$$l = \sqrt{8^2 + 4.5^2}$$

$$= 9.179\,\text{m}$$

$$\text{Surface area} = \frac{1}{2} \times \text{base} \times \text{height} \times \text{no. of triangles}$$

$$= \frac{1}{2} \times 9 \times 9.179 \times 4$$

$$= 165.22\,\text{m}^2$$

EXAMPLE 2.9

Find the total surface area of a hexagonal pyramid of side length 4 m and height 6 m.

A hexagonal base consists of six equilateral triangles (see Fig. 2.8). Therefore to find the area, we find the area of one triangle and multiply it by 6.

Figure 2.8

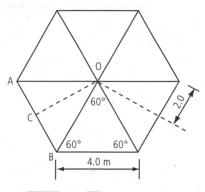

$$\text{Length OC} = \sqrt{4^2 - 2^2} = \sqrt{12} = 3.46$$

$$\text{Area of one base triangle} = \frac{1}{2} \times \text{base} \times \text{height}$$

$$= \frac{1}{2} \times 4 \times 3.46 = 6.92$$

Area of pyramid base = area of one triangle × no. of triangles

$$= 6.92 \times 6$$

$$= 41.52 \, \text{m}^2$$

To find *l* we need to know the length from the centre of the sides to the centre of the base (see Fig. 2.9):

Figure 2.9

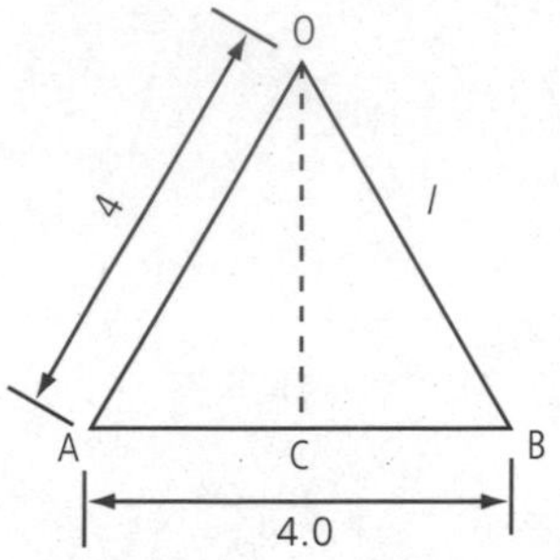

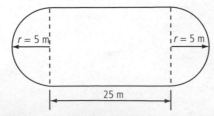

Slant height of side $= \sqrt{6^2 + 3.46^2} = \sqrt{36 + 12} = \sqrt{48} = 6.93$

Surface area of one side $= \dfrac{1}{2} \times$ base × slant height

$$= \frac{1}{2} \times 4 \times 6.93 = 13.96$$

Surface area of pyramid = surface area of one side × no. of sides

$$= 13.96 \times 6$$

$$= 83.16 \, \text{m}^2$$

Total surface area = area of base + surface area

$$= 41.52 + 83.16$$

$$= 124.68 \, \text{m}^2$$

Composite figures

Composite figures are a combination of the different regular shapes that we are already familiar with joined together. In order to calculate the area of composite figures, we must break them down into their known component shapes.

EXAMPLE 2.10

Fig. 2.10 shows a rectangular shape with semicircular ends. Calculate the area of the shape.

Figure 2.10

$r = 5$ m $r = 5$ m

25 m

$$\text{Area of rectangle} = L \times B$$

$$= 25 \times 10 \quad = 250.00$$

$$\text{Area of two semicircles} = \pi r^2$$

$$= 3.142 \times 5^2 = 78.54$$

$$\text{Total} = 328.54 \, m^2$$

EXAMPLE 2.11

Calculate the area of the shape in Fig. 2.11. This composite shape comprises a trapezium and a triangle.

Figure 2.11

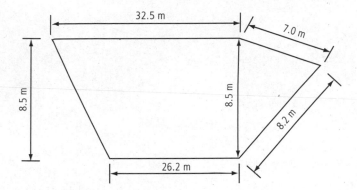

$$\text{Area of trapezium} = (\frac{1}{2} \times \text{sum of parallel sides}) \times (\text{perpendicular distance between them})$$

$$= \frac{1}{2}(32.5 + 26.2) \times 8.5$$

$$= 29.35 \times 8.5 = 249.475 \, m^2$$

$$\text{Area of triangle} = \sqrt{s(s-a)(s-b)(s-c)}$$

where

$$s = \frac{a+b+c}{2}$$

$$s = \frac{7.0 + 8.2 + 8.5}{2}$$

$$= 11.85$$

$$\text{Area of triangle} = \sqrt{(11.85)(11.85 - 7.0)(11.85 - 8.2)(11.85 - 8.5)}$$

$$= \sqrt{702.74}$$

$$= 26.51 \, m^2$$

$$\text{Total area} = 249.475 + 26.51$$

$$= 275.98 \, m^2$$

EXAMPLE 2.12

Calculate the area of the composite shape shown in Fig. 2.12.

Figure 2.12

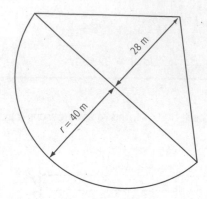

This shape comprises a semicircle and a triangle

$$\text{Area of semicircle} = (\pi r^2)(0.5)$$
$$= (\pi \times 40^2)(0.5)$$
$$= 2513.27\,\text{m}^2$$
$$\text{Area of triangle} = \frac{1}{2} \times \text{base} \times \text{height}$$
$$= (0.5)(80)(28)$$
$$= 1120\,\text{m}^2$$
$$\text{Total area} = 3633.27\,\text{m}^2$$

Volumes

We are already familiar with calculating the volume of such shapes as a prism and a cylinder. From this we know that we need to have three dimensions to give the cubic content of volumes of pyramids and cones.

To calculate the volume of a pyramid or cone we use the following formula:

$$\text{volume} = \frac{1}{3} \times \text{area of base} \times \text{perpendicular height}$$

EXAMPLE 2.13

Calculate the volume of a square pyramid 4 × 4 m with a perpendicular height of 9 m.

$$\text{Volume} = \frac{1}{3} \times \text{area of base} \times \text{perpendicular height}$$
$$= \frac{1}{3} \times 4 \times 4 \times 9$$
$$= 48\,\text{m}^3$$

EXAMPLE 2.14

Calculate the volume of a cone with a base radius of 2 m and a perpendicular height of 5 m.

$$\text{Volume} = \frac{1}{3} \times \text{area of base} \times \text{perpendicular height}$$

$$= \frac{1}{3} \times \pi \times 2 \times 2 \times 5$$

$$= 20.94 \, \text{m}^3$$

Sometimes we need to find the volume and surface areas of cones and pyramids that are not complete, that is the top is cut off. When the top is cut off by a plane parallel to the base of the pyramid or cone, the portion remaining is called the frustum of the pyramid or cone (see Fig. 2.13).

Figure 2.13

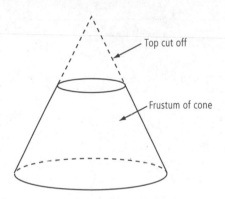

Top cut off

Frustum of cone

EXAMPLE 2.15

Find the volume of the frustum of the cone whose perpendicular height is 8 m and whose radii of the top and base are 2 and 4 m, respectively.

As always, it will help us to sketch the problem, as shown in Fig. 2.14. To calculate the volume of the frustum we need to find the volume of the cone ABC and subtract the volume of the cone Amn from it. To do this we need the total height of the cone ABC and the height of cone Amn. These can be found by using similar triangles.

To find x using similar triangles Amp and mBq

$$\frac{\text{Ap}}{\text{mp}} = \frac{\text{mr}}{\text{Br}}$$

Substituting x for Ap, we have

$$\frac{x}{2} = \frac{8}{2}$$

$$x = \frac{8}{2} \times 2 = 8 \, \text{m}$$

Figure 2.14

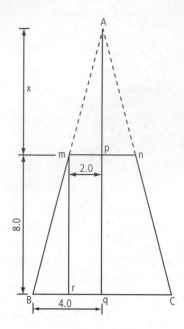

Total height $= 8 + 8 = 16\,\text{m}$

Volume of large cone $= \dfrac{1}{3}\pi r^2 h$

$= \dfrac{1}{3}\pi \times 4^2 \times 16$

$= 268.08\,\text{m}^3$

Volume of small cone $= \dfrac{1}{3}\pi r^2 h$

$= \dfrac{1}{3}\pi \times 2^2 \times 8$

$= 33.51\,\text{m}^3$

Volume of the frustum $= 268.08 - 33.51$

$= 234.57\,\text{m}^3$

EXAMPLE 2.16

Find the total surface area and the volume of a frustum of a square pyramid with a perpendicular height 5 m, length of side of the base 5 m and length of side of the top 3.0 m as shown in Fig. 2.15.

To solve our problem, draw a sketch as shown in Fig. 2.16. We need to find the volume of the large pyramid and the volume of the small pyramid and to subtract one from the other. The first step then is to find the length x.

To find x using similar triangles AEF and FGC

Figure 2.15

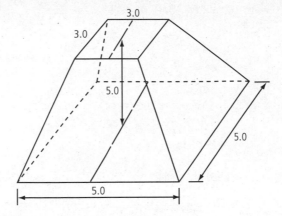

Figure 2.16

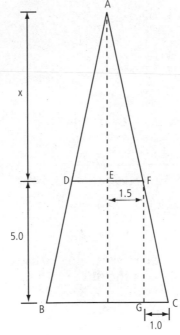

$$\frac{AF}{EF} = \frac{FG}{GC}$$

$$\frac{x}{1.5} = \frac{5}{1}$$

$$x = (5)(1.5) = 7.5$$

Therefore

$$\text{Total height} = 7.5 + 5$$

$$= 12.5 \, \text{m}$$

$$\text{Volume of large pyramid} = \frac{1}{3} \times 5^2 \times 12.5$$

$$= 104.17\,\text{m}^3$$

$$\text{Volume of small pyramid} = \frac{1}{3} \times 3^2 \times 7.5$$

$$= 22.5\,\text{m}^3$$

$$\text{Volume of frustum} = \text{volume of large pyramid} - \text{volume of small pyramid}$$

$$= 104.17 - 22.5$$

$$= 81.67\,\text{m}^3$$

To find the surface area we need to find the slant height FC. Using Pythagoras:

$$FC^2 = 5^2 + 1^2$$

$$FC = \sqrt{5^2 + 1^2}$$

$$= 5.10$$

$$\text{Area of one side} = \frac{3+5}{2} \times 5.1$$

$$= 20.4$$

$$\text{Total surface area} = 20.4 \times 4 + 3^2 + 5^2$$

$$= 115.6\,\text{m}^2$$

See Section 2.2 for further practical applications.

Irregular areas

There are a number of different ways for us to calculate the area of irregular shapes. We are going to look at the three main methods used in the construction industry. They are:

1. Trapezoidal rule

2. Mid-ordinate rule

3. Simpson's rule

TRAPEZOIDAL RULE

This method consists of an application of the rule for finding the area of a trapezoid, i.e. the sum of the two parallel sides divided by 2 multiplied by the perpendicular distance between them.

To apply the rule we draw a base line (see Fig. 2.17a and b) preferably along its longest length, and divide it into a number of equal parts. Ordinates are drawn from each of the points. Each of these sections produced represents a trapezium where the ends are approximately straight lines (the more strips we have, the more accurate the result). The trapezoidal rule can now be applied to each strip.

Referring to Fig. 2.17, each ordinate has been identified by $h_1, h_2, h_3 \ldots h_{10}$ and the distance between them is w.

Figure 2.17

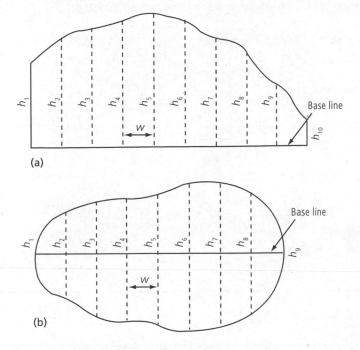

(a)

(b)

The area of the first strip is equal to

$$\frac{h_1 + h_2}{2} \times w$$

The area of the second strip is equal to

$$\frac{h_2 + h_3}{2} \times w$$

thence, the area of last strip is equal to

$$\frac{h_9 + h_{10}}{2} \times w$$

$$\text{Total area} = w\left(\frac{h_1 + h_2}{2} + \frac{h_2 + h_3}{2} + \frac{h_3 + h_4}{2} + \ldots + \frac{h_9 + h_{10}}{2}\right)$$

$$= w\left(\frac{h_1 + 2h_2 + 2h_3 + \ldots + h_{10}}{2}\right)$$

$$= w\left(\frac{h_1 + h_{10}}{2} + h_2 + h_3 + \ldots + h_9\right)$$

Therefore the area of the irregular figure is equal to the width of one strip multiplied by the sum of half the first and last ordinates and all the remaining ordinates.

EXAMPLE 2.17

Find the area of the shape shown in Fig. 2.18 using the trapezoidal rule.

Figure 2.18

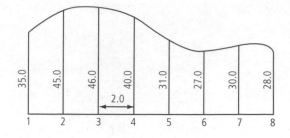

$$\text{Area} = 2\left(\frac{35 + 28}{2} + 45 + 46 + 40 + 31 + 27 + 30\right)$$

$$= 2 \times 250.5$$

$$= 501 \text{ m}^2$$

MID-ORDINATE RULE

This method of calculating irregular areas is similar to the trapezoidal rule in that we divide the area into a number of strips of equal width. Again the more strips we have, the more accurate is the result (see Fig. 2.19). Each strip is assumed to be a trapezium, so therefore the length of the mid-ordinate (the x values in Fig. 2.19) will be the average length of the two parallel sides. The area of each strip will be the length of the mid-ordinate multiplied by the width of the strip, i.e.

wx_1, wx_2, etc.

The total area is the sum of all the strip areas, i.e.

$$wx_1 + wx_2 + wx_3 \ldots + wx_8 = w(x_1 + x_2 + x_3 + \ldots + x_8)$$

Therefore the area of the irregular shape is equal to the width of the strip multiplied by the sum of the mid-ordinates.

Figure 2.19

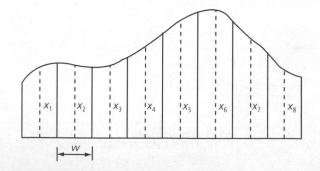

EXAMPLE 2.18

Referring to Fig. 2.19 and given the following information, calculate the area.

$$x_1 = 44.0 \qquad x_2 = 68.5$$
$$x_3 = 67.0 \qquad x_4 = 80.0$$
$$x_5 = 191.0 \qquad x_6 = 120.0$$
$$x_7 = 90.0 \qquad x_8 = 46.0$$
$$w = 30$$

$$\begin{aligned} \text{Area} &= w(x_1 + x_2 + x_3 + \ldots + x_8) \\ &= 30\,(44.0 + 68.5 + 67.0 + 80.0 + 111.0 + 120.0 + 90.0 + 46.0) \\ &= 30 \times 626.5 \\ &= 18\,795 \text{ m}^2 \end{aligned}$$

SIMPSON'S RULE

This rule is generally considered to be the most accurate of the three rules we are considering. Again we divide the area into a number of equal width strips, but there is now one important difference from the previous two methods. The base line *must* be divided into an even number of strips (see Fig. 2.20).

Figure 2.20

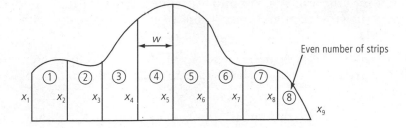

Simpson's rule states that the area of an irregular shape is equal to the width of strip divided by 3 multiplied by the sum of the first and last ordinate, plus the sum of the even numbered ordinates multiplied by 4, plus the sum of the odd numbered ordinates multiplied by 2.

Note: the sum of the odd numbered ordinates does not include the first and the last since they have already been used. Thus

$$\text{area} = \frac{w}{3} \times [(\text{first and last ordinates}) + (4 \times \text{sum of the even ordinates}) + (2 \times \text{sum of the odd ordinates})]$$

EXAMPLE 2.19

Referring to Fig. 2.20, calculate the area given the following data:

$$w = 6 \qquad x_1 = 14.0 \qquad x_2 = 15.0$$
$$x_3 = 15.5 \qquad x_4 = 18.0$$
$$x_5 = 21.0 \qquad x_6 = 19.0$$
$$x_7 = 16.0 \qquad x_8 = 15.5$$
$$x_9 = 0$$

$$\text{Area} = \frac{6}{3} \times [(14.0 + 0) + 4 \times (15.0 + 18.0 + 19.0 + 15.5)] + [2 \times (15.5 + 21.0 + 16.0)]$$

$$= 2 \times [(14.0) + (4 \times 67.5) + (2 \times 52.5)]$$

$$= 2 \times (14.0 + 270 + 105)$$

$$= 2 \times 389$$

$$= 778 \text{ m}^2$$

Volumes of irregular solids

We know from previous work that the volume of a solid with a uniform cross-section is the area of the cross-section multiplied by its length, i.e. a prism or cylinder. If we don't have this uniform cross-section, that is the cross-sections are irregular, then we need to use an alternative method. The method we use is Simpson's rule.

When using Simpson's rule to find irregular areas, we needed to know the lengths of the ordinates. To find the volume of an irregular solid, we need to know the cross-sectional areas. These areas replace the ordinates in the formula.

If we call the cross-sectional areas A_1, A_2, A_3 etc., and these areas are an equal distance d apart, then we state Simpson's rule for the volume of irregular solids as:

$$\text{volume} = \frac{d}{3}\left[\left(\begin{array}{c}\text{first and}\\\text{last area}\end{array}\right) + 4\left(\begin{array}{c}\text{sum of even}\\\text{numbered areas}\end{array}\right) + 2\left(\begin{array}{c}\text{sum of odd}\\\text{numbered areas}\end{array}\right)\right]$$

Note: there must be an odd number of cross-section areas, and the first and last are only used once.

EXAMPLE 2.20

An irregular solid has the following cross-sectional areas taken at 3 m intervals:

section	A_1	A_2	A_3	A_4	A_5	A_6	A_7	A_8	A_9
area	26.4	22.7	21.9	24.6	27.1	28.3	27.5	26.9	26.1

Calculate its volume.

$$\text{Volume} = \frac{d}{3}[(\text{first} + \text{last}) + 4\,(\text{sum of even}) + 2\,(\text{sum of odd})]$$

first: $A_1 = 26.4$

last: $A_9 = 26.1$

        $\overline{52.5}$ 52.5

even: $A_2 = 22.7$

     $A_4 = 24.6$

     $A_6 = 28.3$

     $A_8 = 26.9$

      $\overline{102.5} \times 4 = \;410.0$

odd: $A_3 = 21.9$

    $A_5 = 27.1$

    $A_7 = 27.5$

      $\overline{76.5} \times 2 = \;153.0$

        Total $= \overline{615.5}$

$$\text{Volume} = \left(\frac{3}{3}\right)(615.5)$$

$$= 615.5 \text{ m}^3$$

EXAMPLE 2.21

The trunk of a tree is 10 m long. The cross-sectional areas taken at regular intervals along its length including the ends are 1.13, 1.05, 0.90, 0.79, 0.65 m². Calculate the volume of the trunk.

$$\text{Volume} = \frac{d}{3}[(\text{first} + \text{last}) + 4\,(\text{sum of even}) + 2\,(\text{sum of odd})]$$

$$d = \frac{10}{4} = 2.25$$

first: $A_1 = 1.13$

last: $A_5 = 0.65$

       $\overline{1.78}$ 1.78

even: $A_2 = 1.05$

     $A_4 = 0.79$

      $\overline{1.84} \times 4 = 7.36$

odd: $A_3 = 0.90 \times 2 = 1.80$

        Total $= \overline{10.94}$

$$\text{Volume} = \left(\frac{2.25}{3}\right)(10.94)$$

$$= 8.205 \text{ m}^3$$

See Section 2.2 for further practical applications

Theorems of Pappus

The theorems of Pappus, sometimes referred to as Guldinus's theorems, are used to find the surface area and volume by rotating a line or area about an axis.

If we rotate a triangle about an axis, we will get a cone (see Fig. 2.21). The triangle ABC is rotated about the y–y axis to form a cone.

Figure 2.21

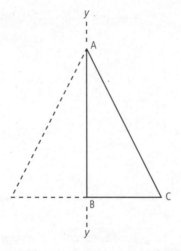

Theorem 1 states that the surface area when a curve (line) is rotated about an axis is equal to the product of the length of the line and the distance travelled by its centroid (centre of gravity).

Referring to Fig. 2.22, the length AC = l, and when rotated around y–y OC = the base radius r. Now the curve AC is a straight line, so therefore its

Figure 2.22

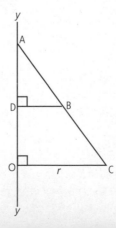

centre of gravity is at the mid-point B and BD will be equal to half of OC; therefore BD $= 0.5r$

When we rotate the line AC about the y–y axis, B will move round a circle whose radius is equal to $0.5r$. The distance moved by B is equal to $= 2\pi(0.5r) = \pi r$.

Applying Pappus's theorem 1, the length of curve l multiplied by the distance travelled by the centroid gives us

$$\text{surface area} = l \times \pi r$$

$$= \pi r l$$

the curved surface of a cone.

EXAMPLE 2.22

An inclined straight line 4 m long rotates about a vertical axis y–y as shown in Fig. 2.23. Calculate the surface area swept out by this rotation.

Figure 2.23

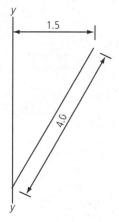

We know that the centre of gravity is the mid-point of the line and that the radius of the circle swept out by this point is half the end radius, giving 0.75 m.

Distance travelled by centroid $= 2\pi r$

$$= 2\pi(0.75)$$

Length of line $l = 4.0$ m

Surface area $=$ length of line $\times$ distanced travelled by centroid

$$= 4 \times 2\pi(0.75)$$

$$= 18.85 \text{ m}^2$$

Theorem 2 states that when an area is rotated about an axis, the volume swept out by the rotating area is equal to the product of the area and the distance travelled by its centroid.

EXAMPLE 2.23

Calculate the volume of the solid swept out by the triangle AOC shown in Fig. 2.24 as it rotates around the axis y–y given that AO $=$ 5 m and OC $=$ 3 m.

Figure 2.24

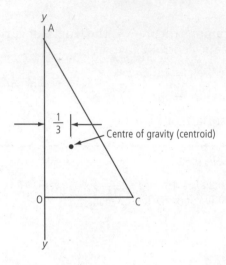

Let AO $= h$, the height of the triangle, and OC $= r$, the radius of the base rotation.

$$\text{Area of triangle} = \frac{\text{base} \times \text{perpendicular height}}{2}$$

$$= \frac{rh}{2}$$

The centroid of the triangle lies on the line a distance OC/3 from the axis y–y, i.e. $r/3$.

Therefore the distance travelled by the centroid will be

$$2\pi \times \frac{1}{3} r = \frac{2\pi r}{3}$$

Volume $=$ area of triangle $\times$ distance travelled by centroid

$$= \frac{rh}{2} \times \frac{2\pi r}{3} = \frac{\pi r^2 h}{3} \text{ (volume of a cone)}$$

$$= \frac{\pi \times 3^2 \times 5}{3} = 47.12 \text{ m}^2$$

EXAMPLE 2.24

Calculate the volume of the ring swept out by a circle as it rotates around an axis (see Fig. 2.25) given that $r = 0.5$ m and $h = 1.5$ m. The centre of the circle is the centroid and the distance it travels is $2\pi r$, where r is equal to h, i.e. $2\pi h$.

$$\text{Area of circle} = \pi r^2$$

$$\text{Volume} = \text{area of circle} \times \text{distance travelled by centroid}$$

Figure 2.25

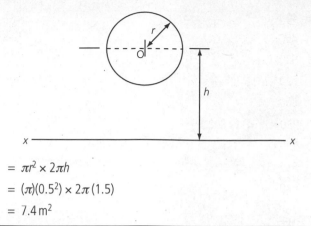

$$= \pi r^2 \times 2\pi h$$
$$= (\pi)(0.5^2) \times 2\pi (1.5)$$
$$= 7.4\,\text{m}^2$$

Practical examples

EXAMPLE 2.25

Find the area of a parcel of land whose boundaries are a straight road on one side and a river on the opposite side given the following ordinates which are spaced at 40 m apart:

| 87.0 | 91.2 | 103.5 | 110.6 | 105.7 | 99.0 | 103.0 |

Using Simpson's rule, then

first + last ordinates = $87.0 + 103.0 = 190.0$

sum of even numbered = 91.2

110.6

99.0

$\overline{300.8 \times 4}$ = 1203.2

sum of odd numbered = 103.5

105.7

$\overline{209.2 \times 2}$ = 418.4

Total = 1811.6

$$\text{Area} = \frac{w}{3}(1811.6) \quad \text{where } w = 40$$

$$\therefore \text{Area} = \frac{40}{3}(1811.6) = 24\,154.7\,\text{m}^2.$$

EXAMPLE 2.26

The cross-sectional areas of a breakwater taken at 75 m intervals are 300, 400, 390, 450, 530, 580, 640, 720, 760, 800, 820. Using Simpson's rule, calculate the volume of stone required for the breakwater.

first + last = 300

820

$\overline{1120}$ = 1120

Sum of even = 400
450
580
720
800
_____
2950 × 4 = 11 800

Sum of odd = 390
530
640
760
_____
2320 × 2 = 4640
_____
Total = 17 560

$$\text{Volume} = \frac{d}{3}(17\,560) \quad \text{where } d = 75$$

$$\therefore \text{Volume} = \frac{75}{3}(17\,560) = 439\,000\,\text{m}^3.$$

EXAMPLE 2.27

Figure 2.26 shows the cross-section of a semicircular concrete retaining wall. Find the volume of concrete required for the wall, if the radius is 8 m.

Figure 2.26

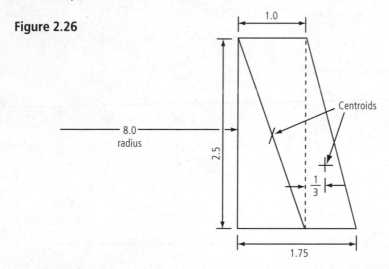

I am going to divide the wall into regular areas as shown in Fig. 2.26 and calculate the volume of each separately.

$$\text{Area of rectangle} = 2.5 \times 1.0 = 2.5$$

$$\text{Distance travelled by centroid} = \frac{2\pi r}{2} \quad \text{where } r = 8.5$$

$$= \pi r$$

$$= \pi \times 8.5$$

$$= 26.7$$

$$\text{Volume} = \text{area} \times \text{distance travelled by centroid}$$

$$= 2.5 \times 26.7$$

$$= 66.75 \text{ m}^3$$

$$\text{Area of triangle} = \frac{\text{base} \times \text{perpendicular height}}{2}$$

$$= \frac{0.75 \times 2.5}{2} = 0.9375 \text{ m}^2$$

$$\text{Distance travelled by centroid} = \frac{2\pi r}{2} \quad \text{where } r = 9.25$$

$$= \pi r$$

$$= \pi \times 9.25 = 29.06 \text{ m}$$

$$\text{Volume} = \text{area} \times \text{distance travelled by centroid}$$

$$= 0.9375 \times 29.06 = 27.24 \text{ m}^3$$

$$\text{Total volume} = 66.75 + 27.24$$

$$= 93.99 \text{ m}^3$$

EXAMPLE 2.30

The cross-section of the excavation for a circular amphitheatre is shown in Fig. 2.27. Given that the length of slope is 10 m, calculate the surface area of the sloping surface.

Figure 2.27

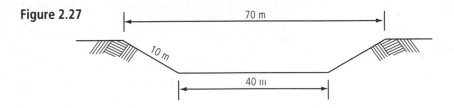

Think of the sloping surface as the rotation of a line whose centroid lies at its mid-point. A sketch of the problem will be useful at this stage (see Fig. 2.28).

Figure 2.28

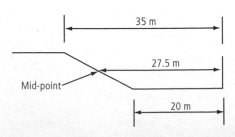

Surface area $=$ length of line $\times$ distance travelled by centroid

$$= 10 \times 2\pi r$$
$$= 10 \times 2\pi \times 27.5$$
$$= 1727.88 \text{ m}^2$$

See Section 2.2 for further practical applications.

2.2 Application exercises

Triangles

Calculate the area of the following triangles:

1 side a = 15 mm   side b = 22 mm   side c = 30 mm

2 side a = 43 mm   side b = 60 mm   side c = 65 mm

3 side a = 105 m   side b = 250 m   side c = 150 m (Ans. in ha)

4 side a = 500 m   side b = 520 m   side c = 570 m (Ans. in ha)

5 side a = 320 m   side b = 280 m   side c = 330 m (Ans. in ha)

6 side a = 75 mm   side b = 40 mm   included angle C = 38°

7 side a = 38 mm   side b = 70 mm   included angle B = 50°

8 side c = 400 m   side b = 450 m   included angle A = 55° (Ans. in ha)

9 side c = 621 m   side a = 587 m   included angle B = 72° (Ans. in ha)

10 side b = 495 m   side c = 567 m   included angle B = 59° (Ans. in ha)

Cylinders

Calculate the curved surface area of the following cylinders:

1 radius = 20 mm   height = 400 mm

2 radius = 15 mm   height = 350 mm

3 radius = 50 mm   height = 850 mm

4 radius = 34 mm   height = 2 m

5 radius = 50 mm   height = 1.2 m

Calculate the total surface area of the following cylinders:

6 radius = 80 mm   height = 200 mm

7 diameter = 2 m   height = 4 m

8 diameter = 3 m   height = 3 m

9 radius = 200 mm   height = 1.8 m

10 radius = 1.2 m   height = 3.0 m

Cones

Find the curved surface area of the following cones:

1 radius = 200 mm   perpendicular height = 600 mm

2 diameter = 300 mm   perpendicular height = 500 mm

3 diameter = 6 m   perpendicular height = 10 m

4 radius = 4 m   perpendicular height = 8 m

Find the total surface area of the following cones:

5 radius = 2 m   perpendicular height = 4 m

6 diameter = 1 m   slant height = 3 m

7 radius = 300 mm   slant height = 900 mm

8 radius = 950 mm   perpendicular height = 1200 mm

9 diameter = 680 mm   slant height = 960 mm

10 radius = 425 mm   slant height = 880 mm

Pyramids

Find the sloping surface area of the following pyramids:

1 height = 200 mm   square base, 100 mm side

2 height = 3 m   hexagonal base, 0.2 m side

3 slant height = 1.5 m   octagonal base, 0.25 m side

4 slant height = 800 mm   hexagonal base, 150 mm side

5 height = 2.5 m   rectangular base, 750 × 500 mm

Find the total surface area of the following pyramids:

6 slant height = 1 m   square base, 0.5 m

7 height = 5 m   rectangular base, 3 m × 2 m sides

8 slant height = 3 m   hexagonal base, 750 mm sides

9 slant height = 660 mm   square base, 250 mm side

10 height = 4.2 m   hexagonal base, 0.75 m side

Composite shapes

Calculate the area of the composite shapes shown in Figs 2.29–2.33.

Figure 2.29

Figure 2.30

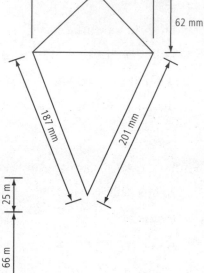

Figure 2.31

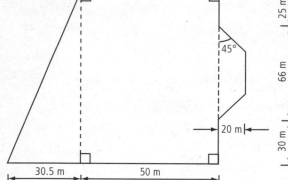

Figure 2.33

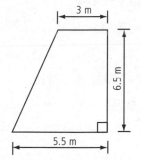

Figure 2.32

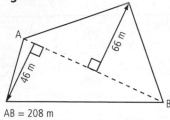

Volumes of cones, pyramids and frustum

1 Find the volume of the following cones:

base radius	perpendicular height
(a) 4.0 m	5.0 m
(b) 2.65 m	3.97 m
(c) 1.89 m	3.15 m

2 Find the volume of the following pyramids:

      base length   breadth   perpendicular height

  (a) 1.5 m   1.5 m   3 m

  (b) 4 m   3 m   7.5 m

  (c) 2.55 m   2.55 m   4.45 m

3 A cone has a slant height of 3.8 m and a base radius of 1.68 m. Calculate its volume.

4 A triangular pyramid has an equilateral triangle base whose length of side is 3.0 m. If the perpendicular height is 4.5 m, calculate its volume and total surface area.

5 A hexagonal pyramid has a perpendicular height of 12 m and the length of the side of the base is 5 m. Calculate its volume.

6 A pyramid has a regular hexagonal base whose sides are 0.75 m. Given that the sloping edge of the pyramid is 3 m, calculate its volume and surface.

7 Calculate the volume of a frustum of a cone given that the radii of the top and bottom surfaces are 2.2 and 3.9 m, respectively, and the perpendicular height between them is 4.0 m.

8 Calculate the volume and total surface area of a frustum whose base measurements are 5 m by 5 m and whose top measures 3 m by 3 m, with a perpendicular height of 6 m.

9 Calculate the volume of a frustum of a cone whose top and bottom diameters are 2 m and 4 m, respectively, given the slant height as 3 m.

10 A rectangular frustum has a base measuring 3 m by 2.5 m and a top measuring 1.5 m by 1.25 m. If the length of its sloping edge is 2.25 m calculate its volume.

Irregular areas and volumes

1 Calculate the area of a piece of land where the offsets have been taken at regular 15 m intervals using:

  (a) the trapezoidal rule
  (b) the mid-ordinate rule

given the following information:

Offset	1	2	3	4	5	6	7	8
Offset length (m)	12.5	14.2	13.6	15.0	14.9	15.2	14.7	14.3

2 At a given point along a river, a series of soundings are taken at equal intervals of 5 m, giving the following results:

Depth (m)   0, 18, 2.2, 2.8, 4.1, 3.3, 2.1, 0

Calculate the cross-sectional area at this point using the mid-ordinate rule.

3 Plot the graph of $y = 12x - 2x^2$ from $x = 0$ to $x = 6$. Find the area enclosed by the curve and the x axis:

(a) using Simpson's rule
(b) between the ordinates 1 and 5 using the trapezoidal and mid-ordinate rule.

4 Find the area of a plot of land given the following ordinate measurements which are 30 m apart and parallel to each other:

90, 98, 95, 115, 132, 121, 102, 93, 88.

5 Figure 2.34 shows the shape of a piece of land a building contractor has bought:

Figure 2.34

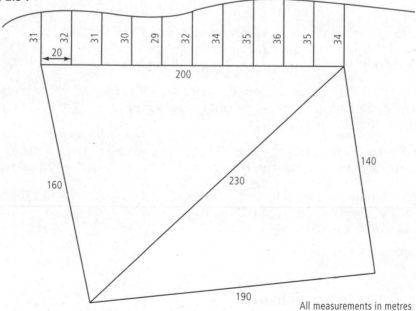

All measurements in metres

(a) Calculate the area of the plot.
(b) How many houses can the contractor build on it allowing 15 houses per hectare?

6 A 30 m high cooling tower is circular in cross-section. The diameters of the tower taken at 5 m intervals commencing at ground level are:

Diameter (m)   20.0, 18.0, 16.0, 14.0, 12.0, 11.0, 10.0

Calculate the volume of the tower.

7 The cross-sectional area of a proposed roadway cutting taken at 20 m intervals are shown in the following table:

Chainage (m)	0	20	40	60	80	100	120	140	160
Area (m²)	48.0	54.2	61.8	72.8	77.4	68.9	65.2	60.1	55.0

Calculate the volume of material to be excavated.

8 Calculate the volume of timber in the bole of a tree if it is 20 m long and has the following cross-sectional areas taken at 5 m intervals: 1.05, 0.95, 0.87, 0.74, 0.69 m².

9 A public sewer 30 m long is to be laid in a horizontal road. The gradient of the sewer is to be 1:60 and the width at the bottom of the trench is 1.1 m. Given that the widths of the trench at the road surface are 1.3 m at the shallow end, 1.5 m at the mid-point and 1.7 m at the deep end, and that the starting depth is 0.9 m, calculate

(a) the volume of soil to be excavated
(b) the volume of soil to be carted away, allowing 18% for bulking.

10 An embankment has to be constructed across a small valley with an angle of repose of 45°. The vertical heights of the embankment taken at 15 m intervals horizontally are 1.9, 2.1, 2.3, 2.6, 2.5, 2.7, 3.3, 3.6, 2.9, 2.7, 2.5, 2.3, 2.1.

Calculate the volume of material required to construct the embankment given that the width at the top is to be 5 m.

Theorems of Pappus

1 A concrete channel has cross-sectional dimensions (mm) as shown in Fig. 2.35. Calculate the volume of concrete required to cast a concrete channel quadrant with an internal diameter of 12 m.

Figure 2.35

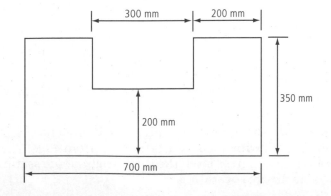

2 A circular concrete kerb has to be cast in-situ. If the internal diameter of the circle is 20 m and the kerb has dimensions as shown in Fig. 2.36 calculate the volume of concrete required to construct the kerb.

Figure 2.36

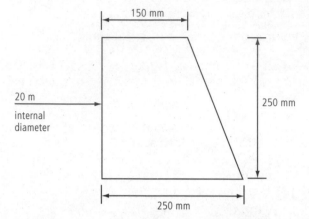

3 The circular timber ribs to a barrel roof subtend an angle of 120° at the centre with a radius of curvature of 50 m. The ribs are of rectangular section measuring 250 × 450 mm. If 10 such ribs are required, calculate the total weight in tonnes if timber weighs 450 kg/m³.

4 Figure 2.37 shows the cross-section of a circular concrete retaining wall. The wall subtends an angle of 140° at its centre with a radius of 20 m. Calculate the volume of concrete required to construct the wall.

Figure 2.37

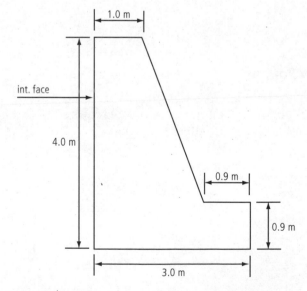

5 A spectator stand to be built at one end of a sports stadium will be semicircular in plan with a radius of 50 m to its front edge. The rake of the seating is 40° to the horizontal. If the total height of the seating area is 20 m, calculate the area of seating.

2.3   Answers to Section 2.2

Triangles

1 157.94 mm^2

2 1253.18 mm^2

3 0.309 ha

4 12.046 ha

5 4.1 ha

6 923.49 m^2

7 1018.84 mm^2

8 7.37 ha

9 17.33 ha

10 12.03 ha

Cylinders

1 50 265.48 mm^2

2 32 986.72 mm^2

3 267 035.38 mm^2

4 0.427 m^2

5 0.377 m^2

6 140 743.35 mm^2

7 31.42 m^2

8 42.41 m^2

9 2.51 m^2

10 31.67 m^2

Cones

1 397 383.53 mm^2

2 245 993.93 mm^2

3 98.4 m^2

4 112.4 m^2

5 40.67 m^2

6 5.5 m^2

7 1 130 973.36 mm^2

8 7 403 152.56 mm^2

9 1 388 583.95 mm^2

10 1 742 405.83 mm^2

Pyramids

1 41 231.06 mm^2

2 1.8 m^2

3 1.5 m^2

4 360 000 mm^2

5 3.15 m^2

6 1.25 m^2

7 31.74 m^2

8 8.21 m^2

9 392 500 mm^2

10 11.02 m^2

Composite shapes

Fig. 2.29   16 111.1 mm^2

Fig. 2.30   12 380 mm^2

Fig. 2.31   8815.25 m^2

Fig. 2.32   11 648 m^2

Fig. 2.33   27.63 m^2

Volumes of cones, pyramids and frustum

1 (a) $83.77 \, \text{m}^3$
   (b) $29.19 \, \text{m}^3$
   (c) $11.78 \, \text{m}^3$

2 (a) $2.25 \, \text{m}^3$
   (b) $30 \, \text{m}^3$
   (c) $9.64 \, \text{m}^3$

3 $10.07 \, \text{m}^3$

4 Volume $= 5.85 \, \text{m}^3$
   Total surface area $= 24.97 \, \text{m}^2$

5 $259.8 \, \text{m}^3$

6 Surface area $= 8.16 \, \text{m}^2$
   Volume $= 1.42 \, \text{m}^3$

7 $119.92 \, \text{m}^3$

8 Total surface area $= 131.32 \, \text{m}^2$
   Volume $= 98 \, \text{m}^3$

9 $20.73 \, \text{m}^3$

10 $8.87 \, \text{m}^3$

Irregular areas and volumes

1 (a) $1515 \, \text{m}^2$
   (b) $1716 \, \text{m}^2$

2 $81.5 \, \text{m}^2$

3 (a) $713.33 \, \text{units}^2$
   (b) $600 \, \text{units}^2$, $700 \, \text{units}$

4 $16\,960 \, \text{m}^2$

5 (a) 3.543 hectares
   (b) 53 houses

6 $5057.96 \, \text{m}^3$

7 $10\,238.67 \, \text{m}^2$

8 $15.627 \, \text{m}^3$

9 (a) $45.1 \, \text{m}^3$
   (b) $53.22 \, \text{m}^3$

10 $3543.5 \, \text{m}^3$

Theorems of Pappus

1 $3.889 \, \text{m}^3$

2 $6.315 \, \text{m}^3$

3 5.3 tonnes

4 $833.87 \, \text{m}^3$

5 $6052.5 \, \text{m}^2$

chapter

3

Trigonometric techniques

Outcomes

At the end of this chapter you should be able to:

- use basic trigonometrical formulae for sine, cosine and tangent
- apply the sine rule
- apply the cosine rule
- understand how to apply different formulae to solving areas of triangles.

3.1 Formulae for right-angled triangles

Trigonometry is the branch of mathematics that deals with the measurement of sides and angles of triangles. In construction we make considerable use of these techniques for such work as

1. setting out – building, roads, roofs
2. surveying – location of points, various calculations
3. frameworks – force diagrams.

In order to do this we must know and understand various techniques and formulae.

We will start with the three basic formulae that we should already be familiar with, but we will re-cap on them before progressing any further.

The three formulae apply only to acute right-angled triangles. Referring then to Fig. 3.1, note that the angles are identified with capital letters and the sides opposite the angles with the same letter but in the lower case. In addition to this, the two sides containing the right angle are called the opposite, i.e. opposite the angle under consideration, and the adjacent, i.e. the side adjacent to the angle under consideration. The third side is known as the hypotenuse.

The basic formulae that we are looking at are as follows:

Figure 3.1

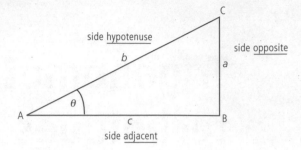

- the tangent of the angle
- the sine of the angle
- the cosine of the angle

What we must remember here is that these are simply the ratios of two sides of our triangle.

Tangent of the angle

Referring to Fig. 3.1, providing angle θ remains constant, the ratio between the side opposite and the side adjacent will be the same no matter how big the right-angled triangle and it is this ratio that we call the tangent of the angle. We say that

$$\text{tangent } \theta = \frac{\text{side opposite}}{\text{side adjacent}}$$

Sine of the angle

In the case of the sine of the angle, the same argument applies but this time the two sides we compare are the side opposite and the hypotenuse. We say that

$$\text{sine } \theta = \frac{\text{side opposite}}{\text{hypotenuse}}$$

Cosine of the angle

With the cosine of the angle we use the side adjacent and the hypotenuse. We say that

$$\text{cosine } \theta = \frac{\text{side adjacent}}{\text{hypotenuse}}$$

Knowing these three formulae allows us to solve various problems involving right-angled triangles.

EXAMPLE 3.1

Find the unknown sides of the following right-angled triangles:

	Angle	Opposite	Adjacent	Hypotenuse
(a)	33	3	–	–
(b)	–	5	15	–
(c)	40	–	6	–
(d)	–	–	8	16
(e)	39	–	–	14
(f)	–	7	–	15

When solving trigonometric problems it is always a good idea to sketch what we know about the problem and show what we are trying to solve.

(a)

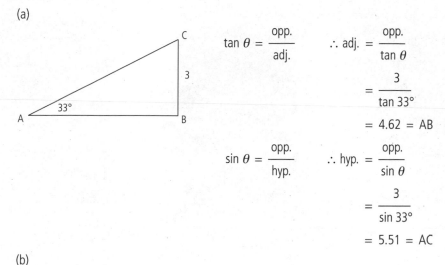

$$\tan \theta = \frac{\text{opp.}}{\text{adj.}} \qquad \therefore \text{adj.} = \frac{\text{opp.}}{\tan \theta}$$

$$= \frac{3}{\tan 33°}$$

$$= 4.62 = AB$$

$$\sin \theta = \frac{\text{opp.}}{\text{hyp.}} \qquad \therefore \text{hyp.} = \frac{\text{opp.}}{\sin \theta}$$

$$= \frac{3}{\sin 33°}$$

$$= 5.51 = AC$$

(b)

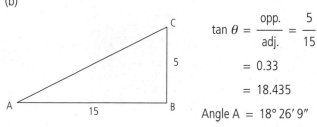

$$\tan \theta = \frac{\text{opp.}}{\text{adj.}} = \frac{5}{15}$$

$$= 0.33$$

$$= 18.435$$

Angle A $= 18° \, 26' \, 9''$

Using Pythagoras to find the hypotenuse we get

$$AC = \sqrt{15^2 + 5^2}$$

$$= 15.81$$

(c)

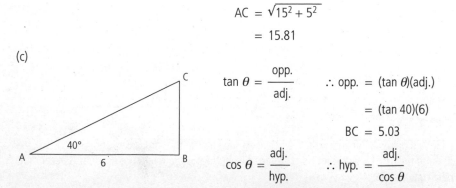

$$\tan \theta = \frac{\text{opp.}}{\text{adj.}} \qquad \therefore \text{opp.} = (\tan \theta)(\text{adj.})$$

$$= (\tan 40)(6)$$

$$BC = 5.03$$

$$\cos \theta = \frac{\text{adj.}}{\text{hyp.}} \qquad \therefore \text{hyp.} = \frac{\text{adj.}}{\cos \theta}$$

$$= \frac{6}{\cos 40}$$

$$AC = 7.83$$

(d)

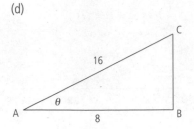

$$\cos \theta = \frac{adj.}{hyp.}$$

$$= \frac{8}{16}$$

$$= 0.5$$

Angle A = 60°

Using Pythagoras to find BC we get

$$BC = \sqrt{16^2 - 8^2}$$

$$= 13.86$$

(e)

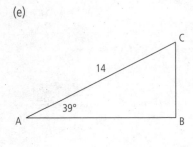

$$\sin \theta = \frac{opp.}{hyp.} \qquad \therefore opp. = (\sin \theta)(hyp.)$$

$$= (\sin 39)(14)$$

$$BC = 8.81$$

$$\cos \theta = \frac{adj.}{hyp.} \qquad \therefore adj. = (\cos \theta)(hyp.)$$

$$= (\cos 39)(14)$$

$$AB = 10.88$$

(f)

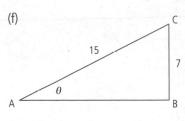

$$\sin \theta = \frac{opp.}{hyp.} = \frac{7}{15}$$

$$= 0.467$$

Angle A = 27° 49′ 5″

Using Pythagoras we get

$$AB = \sqrt{15^2 - 7^2}$$

$$= 13.27$$

EXAMPLE 3.2

A tiled roof requires a minimum pitch of 22.5°. If the span of the roof is 8 m and the rise 2 m, will the slope of the roof be sufficient for the given tiles? What is the minimum rise that is acceptable for this particular roof?

Sketch the given roof.

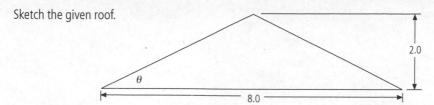

We need to know if θ is 22.5° or more.

Sketch half the roof.

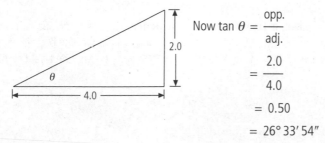

Now tan θ = $\dfrac{\text{opp.}}{\text{adj.}}$

$= \dfrac{2.0}{4.0}$

$= 0.50$

$= 26° 33' 54''$

Since this angle is greater than 22.5° the slope of the roof is acceptable.

The acceptable minimum rise for 22.5°

tan 22.5 $= \dfrac{\text{opp.}}{4.0}$ $\therefore$ opposite = tan 22.5 × 4

minimum rise = 1.66 m

EXAMPLE 3.3

Calculate the height of the mill chimney shown in Fig. 3.2, given that the height of the surveying instrument above ground level is 1.45 m and the angle of elevation is 35°.

Figure 3.2

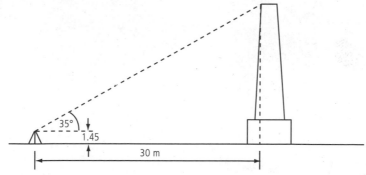

Sketch of the problem: we need to calculate h and remember to add on 1.45 m for the height of the instrument.

tan θ = $\dfrac{\text{opposite}}{\text{adjacent}}$

$\therefore$ opposite = tan θ × adjacent

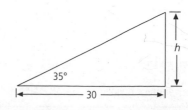

Substitute values:

$$h = \tan 35 \times 30$$
$$= 21.006 \, \text{m}$$
$$\text{Total height} = 21.006 + 1.45$$
$$= 22.456 \, \text{m}$$

EXAMPLE 3.4

A flagstaff stands on the top of a building. From a position 30 m from the building, the angle of depression to the foot of the building is 10° and the angles of elevation to the top and bottom of the flagstaff are 25° and 33°, respectively. Calculate the height of the flagstaff and the total height of the building and flagstaff.

Sketch the problem.

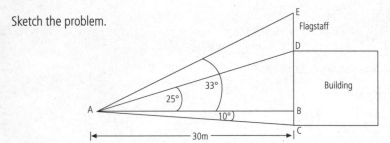

Let DE represent the flagstaff. Then

$$\frac{DB}{AB} = \tan 25$$

$$\therefore DB = \tan 25 \times 30 = 13.989 \, \text{m}$$

Then

$$\frac{EB}{AB} = \tan 33$$

$$EB = \tan 33 \times 30 = 19.482 \, \text{m}$$

Hence

$$EB - DB = DE$$

$$19.482 - 13.989 = 5.493 \, \text{m flagstaff height}$$

$$\tan 10 = \frac{BC}{30}$$

$$BC = \tan 10 \times 30 = 5.29$$

$$\text{Height of building} = 19.482 + 5.29 = 24.772 \, \text{m}$$

See Section 3.3 for further practical applications.

3.2 Formulae for non-right-angled triangles

So far we have only considered triangles that contain a right angle. This allows us to solve all related problems to right-angled triangles by using the trigonometric ratios sine, cosine, tangent and Pythagoras, providing we know at least three details or facts about them.

Not all triangles are right-angled but all triangles do have three sides and three angles and provided we know three facts about them we can solve all related problems. Any triangle can be solved if we are given

- two sides and two angles
- two sides and the contained angle
- two sides and the angle not contained, but in this case there may be two solutions.

We have two methods for solving triangles meeting the above criteria. Which method we choose depends on the information or facts that we know about the triangle.

The two methods are:

- sine rule
- cosine rule.

The sine rule

The sine states that

$$\frac{a}{\sin A} = \frac{b}{\sin B} = \frac{c}{\sin C}$$

This rule can also be written as

$$\frac{\sin A}{a} = \frac{\sin B}{b} = \frac{\sin C}{c}$$

A triangle can be solved using this rule when we know one of the sides and two angles.

EXAMPLE 3.5

In a triangle ABC, $A = 58°$, $B = 65°$ and the length $b = 400$ mm. Find the other two sides.

The third angle can be found:

$$C = 180 - 58 - 65 = 57°$$

Applying the sine rule:

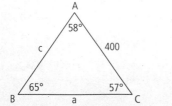

$$\frac{a}{\sin A} = \frac{b}{\sin B} = \frac{c}{\sin C}$$

$$\frac{a}{\sin 58} = \frac{400}{\sin 65} = \frac{c}{\sin 57}$$

$$\text{side } a = \frac{400\,(\sin 58)}{\sin 65} = 374.3 \text{ mm}$$

$$\text{side } c = \frac{400\,(\sin 57)}{\sin 65} = 370.1 \text{ mm}$$

A quick check to make sure we have calculated correctly is that the largest angle should be opposite the largest side and the smallest angle should be opposite the smallest side.

EXAMPLE 3.6

Solve the triangle ABC given that angle A = 79° 36′, a = 195 mm, c = 137 mm.

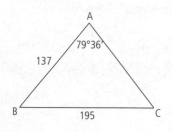

$$\frac{a}{\sin A} = \frac{b}{\sin B} = \frac{c}{\sin C}$$

$$\frac{195}{\sin 79° \, 36'} = \frac{137}{\sin C}$$

$$\sin C = \frac{(137)(\sin 79° \, 36')}{195} = 0.691$$

$$\therefore C = 43° \, 42' \, 40''$$

$$B = 180 - (79° \, 36' + 43° \, 42' \, 40'')$$

$$= 56° \, 41' \, 20''$$

Side b:

$$\frac{a}{\sin A} = \frac{b}{\sin B}$$

$$b = \frac{(a)(\sin B)}{\sin A} = \frac{(195)(\sin 56° \, 41' \, 20'')}{\sin 79° \, 36'}$$

$$= 171.15 \text{ mm}$$

EXAMPLE 3.7

A triangle ABC has a base BC = 67 m, angle B = 41° 20′ 25″ and angle C = 64° 39′ 45″. Find the lengths of AB and BC.

Sketching the problem we get:

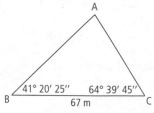

$$\therefore \text{ Angle A} = 180° - (41° \, 20' \, 25'' + 64° \, 39' \, 45'')$$

$$= 73° \, 59' \, 50''$$

Apply the sine rule:

$$\frac{a}{\sin A} = \frac{b}{\sin B}$$

$$\frac{67}{\sin 73°\,59'\,50''} = \frac{b}{\sin 41°\,20'\,25''}$$

$$\therefore b = \frac{(67)(\sin 41°\,20'\,25'')}{\sin 73°\,59'\,50''}$$

$$b = 46.04 \text{ m}$$

$$\frac{a}{\sin A} = \frac{c}{\sin C}$$

$$\frac{67}{\sin 73°\,59'\,50''} = \frac{c}{\sin 64°\,39'\,45''}$$

$$c = \frac{(67)\,(\sin 64°\,39'\,45'')}{\sin 73°\,59'\,50''}$$

$$c = 62.996$$

EXAMPLE 3.8

To find the height of a hill, a horizontal line 180 m long was set out some distance away from the hill but in line with the top T, i.e. in the same vertical plane (see Fig. 3.3). From R and S the angles of elevations to T were found to be 35° 30' and 43° 20', respectively. Calculate the height of the hill.

Figure 3.3

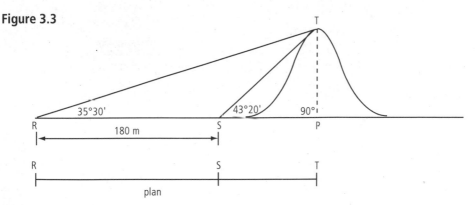

With reference to Fig. 3.3:

$$\text{angle RTS} = 43°\,20' - 35°\,30'$$

$$= 7°\,50'$$

Applying the sine rule:

$$\frac{ST}{\sin 35°\,30'} = \frac{180}{\sin 7°\,50'}$$

$$ST = \frac{(180)(\sin 35° \, 30')}{\sin 7° \, 50'}$$

$$= 766.931$$

In triangle SPT:

$$\frac{TP}{ST} = \frac{TP}{766.931} = \sin 43° \, 20'$$

$$\therefore TP = (766.931)(43° \, 20')$$

$$TP = 526.3 \, m, \text{ the height of the hill}$$

EXAMPLE 3.9

The boundaries of a plot of land form a triangle ABC. If AB is 650 m and angles A and B are 48° and 68°, respectively, find the shortest distance from C to AB.

Sketch the problem:

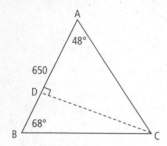

The shortest distance is a line at right angles from AB to C, i.e. DC.

$$\text{Angle } C = 180 - (48 + 68) = 64°$$

Applying the sine rule:

$$\frac{BC}{\sin A} = \frac{AB}{\sin C}$$

$$\frac{BC}{\sin 48} = \frac{650}{\sin 64}$$

$$BC = \frac{650 \, (\sin 48)}{\sin 64}$$

$$BC = 537.436 \, m$$

Now from triangle DBC, we can calculate the length DC:

$$\sin 68 = \frac{DC}{BC}$$

$$\therefore DC = (\sin 68)(537.436)$$

$$DC = 498.302 \, m$$

See Section 3.3 for further practical applications.

The cosine rule

The cosine rule states that

$$a^2 = b^2 + c^2 - 2bc \cos A$$

This rule can also be written as

$$b^2 = a^2 + c^2 - 2ac \cos B$$

or

$$c^2 = a^2 + b^2 - 2ab \cos C$$

A triangle can be solved using this rule when we know the lengths of two sides and the included angle or when we know the three sides.

If we know the two sides and included angle then we use the rule in the above form, but when we know the three sides we need to rearrange the rule to make the cosine of the angle the subject, i.e.

$$\cos A = \frac{b^2 + c^2 - a^2}{2bc}$$

or

$$\cos B = \frac{a^2 + c^2 - b^2}{2ac}$$

or

$$\cos C = \frac{a^2 + b^2 - c^2}{2ab}$$

EXAMPLE 3.10

Given that the two sides and included angle of a triangle are, respectively, 8 m, 12 m and 54°, find the length of the third side.

Sketch the problem:

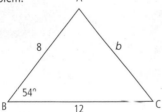

Apply the cosine rule:

$$b^2 = a^2 + c^2 - 2ac \cos B$$
$$b^2 = 12^2 + 8^2 - (2)(12)(8) \cos 54$$
$$= 95.15$$
$$\therefore b = \sqrt{95.15}$$
$$b = 9.75$$

EXAMPLE 3.11

Find the length of the third side of a triangle given that two sides are 68 m and 50 m and the included angle = 120°.

Sketch the problem:

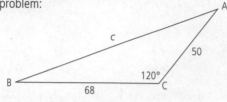

$$c^2 = a^2 + b^2 - 2ab \cos C$$

$$c^2 = 68^2 + 50^2 - (2)(68)(50) \cos 120°$$

$$= 10\,524$$

$$c = \sqrt{10\,524}$$

$$c = 102.\,59$$

EXAMPLE 3.12

Solve the triangle RST given that angle R = 117°, RS = 35 and RT = 18.

Sketch the problem:

$$r^2 = s^2 + t^2 - 2\,st \cos R$$

$$= 18^2 + 35^2 - (2)(18)(35) \cos 117$$

$$= 2121.03$$

$$r = \sqrt{2121.03}$$

$$r = 46.05$$

Now that we know the third side, we can use the cosine rule to solve the remaining angles:

$$\cos S = \frac{t^2 + r^2 - s^2}{2tr}$$

$$\cos S = \frac{35^2 + 46.05^2 - 18^2}{(2)(35)(46.05)}$$

$$\cos S = 0.9374$$

$$S = 20.386°$$

$$S = 20° 23' 10''$$

$$R = 180 - (117 + 20.386)$$

$$= 42.619°$$

$$= 42° 36' 50''$$

It is worth noting that we could have used the sine rule to solve the angles.

EXAMPLE 3.13

A parallelogram has sides of length 2 m and 3 m with an included angle of 65°. Calculate the length of the two diagonals.

Sketch the problem:

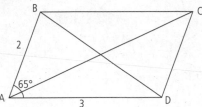

The angles contained in the parallelogram are

$$D = 115°$$

$$C = 65°$$

$$B = 115°$$

In triangle ABD:

$$BD^2 = AD^2 + AB^2 - (2)(AD)(AB) \cos A$$

$$= 3^2 + 2^2 - (2)(3)(2) \cos 65$$

$$BD^2 = 7.92$$

$$BD = 2.814$$

In triangle ACD:

$$AC^2 = CD^2 + AD^2 - (2)(CD)(AD) \cos D$$

$$= 2^2 + 3^2 - (2)(2)(3) \cos 115$$

$$AC^2 = 18.07$$

$$AC = 4.25$$

EXAMPLE 3.14

In Fig. 3.4, RSU is in the horizontal plane and TU, 18.5 m high, is in the vertical plane. Calculate the lengths RU, SU and RS.

Since RSU is in the horizontal plane and UT is vertical, RTU and STU are right-angled triangles. Using basic trigonometric formulae we can calculate the lengths RU and SU.

Figure 3.4

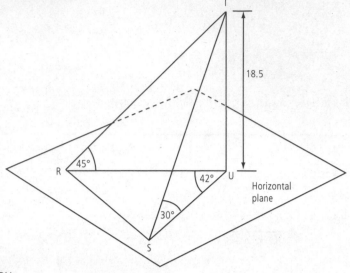

To calculate RU.

Sketch the problem:

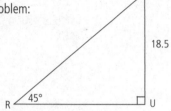

From the sketch we can see RTU is a 45° triangle. Then RU = TU = 18.5 m.

To calculate SU.

Sketch the problem:

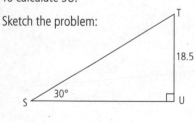

$$\text{Tan}\ \theta = \frac{\text{opp.}}{\text{adj.}}$$

$$\text{adj.} = \frac{\text{opp.}}{\tan \theta}$$

$$\text{then SU} = \frac{18.5}{\tan 30}$$

$$\text{SU} = 32.04\,\text{m}$$

To calculate RS.

Sketch the problem:

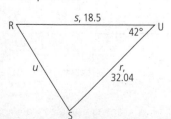

Cosine rule states

$$u^2 = r^2 + s^2 - 2rs \cos U$$

$$u^2 = 32.04^2 + 18.5^2 - (2)(32.04)(18.5) \cos U$$

$$u^2 = 487.83$$

$$u = 22.09$$

$$\therefore \text{RS} = 22.08\,\text{m}$$

See Section 3.3 for further practical applications

To find the area of any triangle

We are already familiar with finding the area of a triangle using the 'half the base multiplied by perpendicular height' formula. Sometimes the data we have is not always as simple or ready to use as we require for this formula. To overcome this problem we have two alternatives available to us.

1. If we know two sides and the included angle we can use the following formula:

$$\frac{1}{2} ab \sin C$$

where a and b are the two known sides and C is the included angle.

2. If we know the lengths of the three sides we can use the following formula:

$$\sqrt{[s(s-a)(s-b)(s-c)]}$$

where a, b and c are the known sides and

$$s = \frac{a+b+c}{2}$$

i.e. half the perimeter of the triangle.

The use of both these formulae is demonstrated in the following examples.

EXAMPLE 3.15

Calculate the area of the following triangles using the $^1/_2$ ab sin C formula:

(i) $a = 85\,m$ $b = 120\,m$ C $= 25°$

(ii) $x = 19.6\,m$ $y = 16.7\,m$ Z $= 79°$

(iii) $d = 210\,mm$ $e = 493\,mm$ F $= 67°\,40'$

(i)   Area $= \dfrac{1}{2} ab \sin C$

   $= (\dfrac{1}{2})(85)(120) \sin 25°$

   Area $= 2155.35\,m^2$

(ii)   Area $= \dfrac{1}{2} xy \sin Z$

   $= (\dfrac{1}{2})(19.6)(16.17) \sin 79°$

   Area $= 160.65\,m^2$

(iii)   Area $= \dfrac{1}{2}\, de \sin F$

$$= (\tfrac{1}{2})(210)(493) \sin 67°\, 40'$$

Area $= 47\,882\,\text{mm}^2$

EXAMPLE 3.16

Calculate the area of the following triangles using the formula $\sqrt{s(s-a)(s-b)(s-c)}$:

(i) $a = 10\,\text{m}$ $b = 18\,\text{m}$ $c = 27\,\text{m}$

(ii) $x = 46\,\text{mm}$ $y = 39\,\text{mm}$ $z = 63\,\text{mm}$

(iii) $d = 127\,\text{m}$ $e = 381\,\text{mm}$ $f = 471\,\text{mm}$

(i)   Area $= \sqrt{s(s-a)(s-b)(s-c)}$

$$s = \frac{10 + 18 + 27}{2} = 27.5$$

Area $= \sqrt{(27.5)(27.5 - 10)(27.5 - 18)(27.5 - 27)}$

$\quad = \sqrt{(27.5)(17.5)(9.5)(0.5)}$

Area $= 47.81\,\text{m}^2$

(ii) Area $= \sqrt{s(s-x)(s-y)(s-z)}$

$$s = \frac{46 + 39 + 63}{2} = 74$$

Area $= \sqrt{(74)(74 - 46)(74 - 39)(74 - 63)}$

$\quad = \sqrt{(74)(28)(35)(11)}$

Area $= 893.15\,\text{mm}^2$

(iii) Area $= \sqrt{s(s-d)(s-e)(s-f)}$

$$s = \frac{127 + 381 + 471}{2} = 489.5$$

Area $= \sqrt{(489.5)(489.5 - 127)(489.5 - 381)(489.5 - 471)}$

$\quad = \sqrt{(489.5)(362.5)(108.5)(18.5)}$

Area $= 18\,872.57\,\text{m}^2$

EXAMPLE 3.17

A quadrilateral has the following dimensions as the result of a chain survey: AB $= 1400\,\text{m}$, BC $= 1300\,\text{m}$, CD $= 1700\,\text{m}$, DA $= 1600\,\text{m}$ and the diagonal BD $= 1900\,\text{m}$.

Calculate the area of the quadrilateral in hectares.

Sketch the problem:

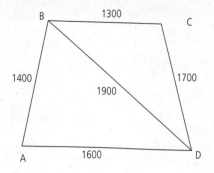

From the diagram we can see that the quadrilateral consists of two triangles and that we know the lengths of sides.

Using the 's' formula:

$$\text{area} = \sqrt{s(s-a)(s-b)(s-c)}$$

Triangle ABD:

$$a = 1900$$

$$b = 1600$$

$$\underline{d = 1400}$$

$$4900$$

$$s = \frac{a+b+c}{2} = \frac{4900}{2} = 2450$$

$$\text{Area} = \sqrt{(2450)(2450-1900)(2450-1600)(2450-1400)}$$

$$= \sqrt{(2450)(550)(850)(1050)}$$

$$= 10^2 \times 5^2 \times 7 \sqrt{11 \times 17 \times 21}$$

$$= 1\,096\,651.15\,m^2$$

$$= 109.67 \text{ hectares} \quad (10\,000\,m^2 = 1\text{ ha})$$

EXAMPLE 3.18

From a point A sightings were taken to other points B and C and the following results obtained:

AB = 196.5   AC = 145.8   angle BAC = 60°30'

Calculate the area of ABC.

Sketch the problem:

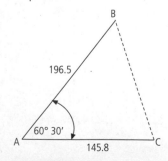

From the diagram we can see that we have two sides and included angle of the triangle.

Using the $^1/_2$ *ab* sin C formula:

$$\text{Area} = \frac{1}{2} bc \sin A$$

$$= (\frac{1}{2})(145.8)(196.5)(\sin 60° \ 30')$$

$$= 12\,467.71 \ m^2$$

See Section 3.3. for further practical applications.

3.3 Practical applications

1 Find the unknown sides in the following right-angled triangles.

	Angle	Opposite	Adjacent	Hypotenuse
(a)	46	–	–	24
(b)	–	21	19	–
(c)	21	–	10.5	–
(d)	–	14.6	–	19.79
(e)	38.5	–	15.86	–

2 A ladder resting against a wall should conform to the ratio of one horizontal to four verticals for safety reasons. Calculate the angle between the foot of the ladder and the ground.

3 A slipway on a beach slopes at an angle of 10° with the horizontal. What will be the length of the slipway when covered by the sea when the water is 5 m deep?

4 The diagram shows a force R inclined at 42° to the horizontal. If the horizontal component is 500 N, calculate the vertical component as shown.

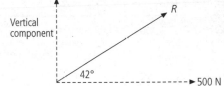

5 Calculate the area of the roof shown on plan in Fig. 3.5. Roof pitch is 40°, units are metres.

Figure 3.5

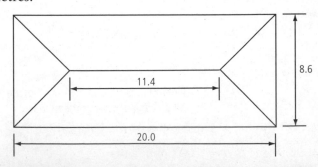

6 Fig. 3.6 shows a concrete retaining wall against an earth bank. Triangle ABC shows the section of the earth supported by the wall. Given that the earth weighs 1.3 tonnes per cubic metre, calculate the weight of earth supported by 1 m run of the wall.

Figure 3.6

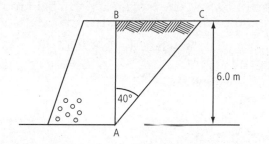

7 An embankment is 4 m high and 7 m wide at the top. If the sides slope 42° to the horizontal, find the number of cubic metres of earth per metre run of the embankment.

8 The section of a north light roof is shown in Fig. 3.7. Find the pitch of the front section of the roof and the span. Units are metres.

Figure 3.7

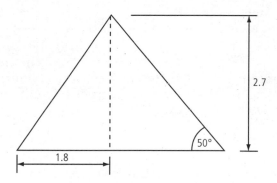

Sine rule

Solve the following triangles:

1 (i)   A = 38°   C = 69°   a = 6.8 m

   (ii)   A = 72°   B = 33°   b = 4 m

   (iii)   B = 62° 15′   C = 48° 20′   c = 8.9 m

   (iv)   a = 8.5 m   b = 12 m   A = 40°

   (v)   b = 145 m   c = 80 m   B = 58°

   (vi)   a = 92.5 m   c = 60 m   C = 38° 30′

   (vii)   C = 102°   c = 849 m   b = 572 m

   (viii) a = 129 m   c = 99 m   A = 74° 35′

   (ix)   C = 48° 40′   c = 45.9 m   B = 53° 25′

   (x)   b = 95.97 m   a = 102.56 m   B = 67°

2 When the top of a pylon P is observed from the two ends of a survey line MN it is found that angle PMN = 363° and angle PNM = 41°. Given that MN is 107 m long, calculate the lengths PM and PN and the horizontal distance from N to the base of the pylon.

3 A roof has a span of 10.5 m. The roof has a north light roof truss, one side sloping at 25° and the other at 55°. Calculate the length of each roof slope.

4 Fig. 3.8 shows a triangle ABC where angle B = 30°, BC = 25 m and AC = 13 m. If BA is extended to D so that AC = CD calculate angles BAC and ACD and also the length of side BD.

Figure 3.8

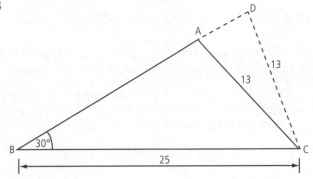

5 A jib crane has a vertical volumn PQ 6 m in height, an inclined jib QR 10 m long and supporting strut PR. The angle between column and inclined jib is 115°. If the inclined strut forms an angle of 40° with the column, calculate the length of the strut.

Cosine rule

1 Solve the following triangles given two sides and the included angle:

(i) $a = 90$ mm $b = 120$ mm   C = 50°

(ii) $x = 48$ m $y = 101$ m   Z = 75°

(iii) $d = 150.6$ m $e = 247.4$ m   F = 67° 35′

(iv) $b = 127.25$ m $c = 104.55$ m   A = 102° 30′ 30″

(v) $r = 59.79$ m $t = 82.31$ m   S = 43° 21′ 29″

2 Solve the following triangles given the three sides:

(i) $a = 85$ m $b = 67$ m $c = 105$ m

(ii) $x = 17.87$ m $y = 21.23$ m $z = 19.39$ m

(iii) $d = 129$ mm $e = 153$ mm $f = 68$ mm

(iv) $a = 91.05$ m $b = 83.76$ m $c = 108.21$ m

(v) $r = 17.68$ m $s = 19.11$ m $t = 23.91$ m

3 Find the angles A, B and C of a triangle given that the lengths $a = 3.5$ m, $b = 4.2$ m and $c = 6.4$ m

4 In a four-sided polygon RSTU, RS = 150 m, ST = 117 m, TU = 195 m, angle RST = 98° 30′ and angle STU = 104° 20′. Find the lengths of the remaining sides and angles.

5 On a land survey the distance between two points A and B cannot be measured directly because of a small wood. From a third point, AC and BC can be measured and are found to be 145 m and 210 m, respectively, and angle BCA = 39°. Find the distance between A and B.

Area of triangles

1 Calculate the area of the following triangles using the $\frac{1}{2}ab$ sin C formula:

(i) a = 42 m b = 63 m   C = 40°

(ii) d = 110 mm e = 74 mm   F = 55° 20′

(iii) r = 89.6 m s = 97.3 m   T = 109° 25′ 30″

(iv) x = 48.65 m y = 36.71 m   Z = 37° 11′ 25″

(v) e = 58 mm f = 71 mm   G = 42° 30′

2 Calculate the area of the following triangles using the 's' formula.
$\sqrt{s\,(s-a)(s-b)(s-c)}$:

(i) a = 9 m b = 11 m c = 14 m

(ii) d = 114 m e = 121 m f = 159 m

(iii) x = 81 mm y = 70 mm z = 106 mm

(iv) r = 49.7 m s = 61.3 m t = 85.2 m

(v) e = 401.55 m f = 472.15 m g = 433.85 m

3 A plot of land has the following dimensions:

AB = 83 m   BC = 90 m   CD = 92 m   DA = 120 m

BD = 146 m

A contractor wishes to develop the site for domestic use. If 15% of the area will be taken up by access roads, how many houses can be built allowing 195 m² per house plot?

4 A theodolite survey has been carried out to find the size of a farmer's field. Calculate the area of the field in hectares if AB = 250 m, BC = 270 m, CD = 270 m, DA = 360 m, angle ABC = 98° 20′ and angle ADC = 70° 30′.

5 The results of a chain survey are as follows

AB = 102 m   BC = 140 m   CD = 124 m   DE = 112 m

EA = 228 m   AD = 270 m   BD = 224 m

Calculate the area.

Further practical applications

1 The survey details shown in Fig. 3.9 are found to be incomplete. Calculate the length of side CD and angle ADC.

Figure 3.9

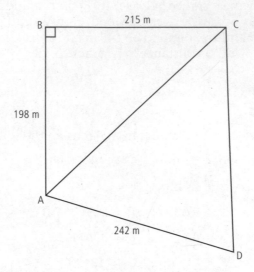

2 In Fig. 3.10, M is a control on a survey. MN and PO are parallel to each other. Determine the angles MPO, MNO, NOP and the length MO.

Figure 3.10

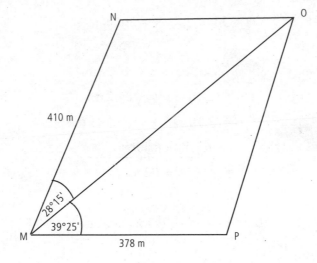

3 Fig. 3.11 shows the plan of a piece of land. Find the lengths EG and FH.

4 The height of a mill chimney due for demolition has to be established. The chimney stands on ground that slopes away with an even gradient of 12° 30′ to the horizontal. From two points A and B in line with the chimney and on a horizontal plane, the angles of elevation to the top of the chimney are measured. From A the angle of elevation measures 39° 32′ 45″, and from B the angle measures 48° 45′ 35″. Calculate the height of the chimney given that A and B are 60 m apart.

Figure 3.11

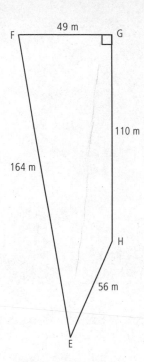

5 Fig. 3.12 represents the access road to a cement works. The cement company is considering constructing a new road from junction A to the cement works at C. Calculate the distance that would be saved by the new route and how many degrees south of east the new road would be from A.

Figure 3.12

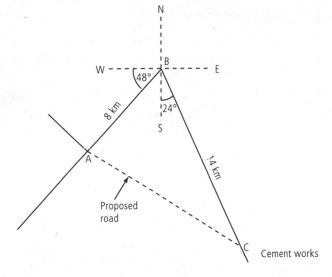

6 Referring to Fig. 3.13, calculate the length of AC.

7 A plot of land ABCDE has been surveyed for a building contractor and the following lengths obtained:

AB = 90 m   BC = 110 m   CD = 90 m   DE = 80 m

EA = 180 m   BE = 162 m   CE = 140 m

Figure 3.13

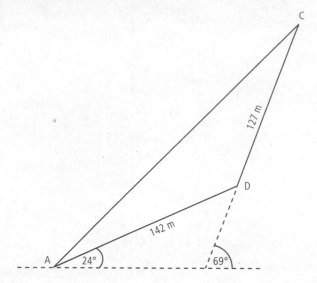

The contractor needs to know the area in square metres and how many dwellings can be built on it at a density of 25 dwellings per hectare.

8 A survey of field produces the following data:

AB = 104.5 m   BC = 97.25 m

CD = 86.75 m   DA = 113.2 m

angle BAC = 87° 15'   angle BCD = 92° 22'

Calculate the area of the field.

9 The three main lines of a chain survey form a triangle XYZ. If XY = 75.65 m, YZ = 83.20 m and ZX = 69.85 m, find the area enclosed by the three sides.

10 Calculate the area of land shown in Fig. 3.14 which a contractor plans to develop. It is the contractor's intention to build domestic dwellings on the land. He estimates that 18% of the land will be required for access roads. If

Figure 3.14

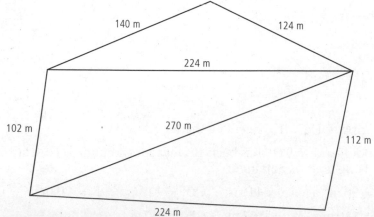

he allows an average of 165 m² for each dwelling, how many dwellings will he be able to build?

Answers to Section 3.3

1 (a) 17.26, 16.67
  (b) 28.32
  (c) 4.03, 11.25
  (d) 13.36
  (e) 12.62, 20.27

2 75.96°

3 28.79 m

4 450.2 N

5 224.53 m²

6 3.272 tonnes

7 45.77 m³

8 Pitch = 56.31°
  Span = 4.066 m

Sine rule

1 (i) $B = 73°$ $b = 10.56\,\text{m}$ $c = 10.31\,\text{m}$
   (ii) $C = 105°$ $a = 6.98\,\text{m}$ $c = 7.09\,\text{m}$
   (iii) $A = 69°\,25'$ $b = 10.54\,\text{m}$ $a = 11.15\,\text{m}$
   (iv) $B = 65°\,9'$ $C = 74°\,51'$ $c = 12.76\,\text{m}$
   (v) $A = 94°\,6'$ $C = 27°\,54'$ $a = 170.54\,\text{m}$
   (vi) $A = 73°\,41'$ $B = 69°\,49'$ $b = 89.25\,\text{m}$
   (vii) $A = 36°\,47'$ $B = 41°\,13'$ $a = 519.64\,\text{m}$
   (viii) $B = 57°\,42'$ $C = 47°\,43'$ $b = 113.11\,\text{m}$
   (ix) $A = 77°\,55'$ $a = 59.77\,\text{m}$ $b = 49.09\,\text{m}$
   (x) $A = 79°\,39'$ $C = 33°\,21'$ $c = 57.32\,\text{m}$

2 PN = 805.44 m   PM = 721.62 m   Distance = 544.61 m

3 4.51 m, 8.73 m

4 BAC = 105° 56′ 33″   ACD = 31° 53′ 6″   BD = 25.22 m

5 14.1 m

Cosine rule

1 (i) $c = 92.82\,\text{mm}$ $A = 47°\,58'\,05''$ $B = 82°\,01'\,55''$
   (ii) $z = 99.98\,\text{mm}$ $X = 27°\,37'\,44''$ $Y = 77°\,22'\,16''$
   (iii) $f = 235.52\,\text{mm}$ $D = 36°\,14'\,11''$ $E = 76°\,10'\,49''$
   (iv) $a = 181.34\,\text{mm}$ $B = 43°\,14'\,30''$ $C = 34°\,15'$
   (v) $s = 56.51\,\text{mm}$ $R = 46°\,35'\,08''$ $T = 90°\,03'\,23''$

2 (i) $A = 53°\,54'\,18''$ $B = 39°\,33'\,47''$ $C = 86°\,31'\,55''$
   (ii) $X = 51°\,57'\,29''$ $Y = 69°\,19'\,55''$ $Z = 58°\,42'\,36''$
   (iii) $D = 56°\,48'\,20''$ $E = 97°\,01'\,10''$ $F = 26°\,10'\,30''$
   (iv) $A = 54°\,51'\,17''$ $B = 48°\,47'\,1''$ $C = 76°\,21'\,42''$
   (v) $R = 46°\,54'\,33''$ $S = 52°\,07'\,25''$ $T = 80°\,58'\,03''$

3 Angle A = 30° 26′ 22″
   Angle B = 37° 26′ 30″
   Angle C = 112° 07′ 08″

4 RU = 191.8 m
 angle TUR = 63° 26′ 55″
 angle SRU = 93° 42′ 41″

5 AB = 133.4 m

Area of triangles

1 (i)   850.4 m²
  (ii)   3347.47 mm²
  (iii) 4110.91 m²
  (iv) 539.77 m²
  (v)   1391 mm²

2 (i)   49.48 m²
  (ii)   6871.8 m²
  (iii) 2834.44 mm²
  (iv) 1501.33 m²
  (v)   81 158.37 m²

3 39 houses

4 7.92 hectares

5 31 400 m²

Further practical applications

1 CD = 284.93 m
 ADC = 67° 25′ 04″

2 MPO = 112° 20′
 MNO = 124° 32′ 47″
 NOP = 55° 27′ 13″
 MO = 738.71 m

3 EG = 163.12 m
 FH = 120.42 m

4 103. 08 m

5 9 km
 31° 46′ 48″

6 248.59 m

7 18 218.66 m²
 46 dwellings

8 10 122.51 m²

9 2476.75 m²

10 136 dwellings

chapter

4

Statistical techniques

4.1 Pictorial representation

Statistics is an area of mathematics that can be used in all walks of life and may be summarised as the collection, presentation and use of certain facts.

In order to make sound use of statistics we must follow a set procedure, which can be described as:

- collection – sampling
- presentation – diagrams
- interpretation – trends
- analysis – testing

of data for planning and future trends.

Collection of data

The 'new' data can be collected in a number of ways, depending on the type of data we are looking for. Possible sources of data could include question-naires, published information, observation of a particular sequence of events, etc.

To be absolutely certain of our data we would need to consider the total event taking place, which we refer to as the 'population'. This is very often time-consuming or not practical. For example, if we wanted to check that a new batch of bricks met the designed strength specification, we could hardly test the entire population of bricks since we would destroy all the bricks in the batch. In order to make it practical, we would take bricks from various parts of the batch to give us a fair cross-section of the bricks and test them. This is known as sampling and we say we are taking a sample of the population.

To record the data we have collected we use a tally diagram.

EXAMPLE 4.1

We wish to know the total number of vehicles crossing a road bridge at hourly intervals between 06.00 and 18.00 hours on a particular day, i.e. 12 hours.

The raw data will be recorded on a tally diagram (see Fig. 4.1).

Figure 4.1

Tally diagram		
Number of vehicles using bridge on Wednesday between 0600 and 1800		
1 hour periods		Frequency
0600 – 0700	~~IIII~~ III	8
0700 – 0800	~~IIII~~ ~~IIII~~ I	11
0800 – 0900	~~IIII~~ ~~IIII~~ IIII	14
0900 – 1000	~~IIII~~ ~~IIII~~ ~~IIII~~ III	18
1000 – 1100	~~IIII~~ ~~IIII~~ ~~IIII~~ ~~IIII~~ ~~IIII~~	25
1100 – 1200	~~IIII~~ ~~IIII~~ ~~IIII~~ ~~IIII~~ ~~IIII~~ II	27
1200 – 1300	~~IIII~~ ~~IIII~~ ~~IIII~~ ~~IIII~~ ~~IIII~~ IIII	29
1300 – 1400	~~IIII~~ ~~IIII~~ ~~IIII~~ ~~IIII~~ ~~IIII~~ I	26
1400 – 1500	~~IIII~~ ~~IIII~~ ~~IIII~~ ~~IIII~~ ~~IIII~~ II	27
1500 – 1600	~~IIII~~ ~~IIII~~ ~~IIII~~ ~~IIII~~ I	21
1600 – 1700	~~IIII~~ ~~IIII~~ ~~IIII~~ II	17
1700 – 1800	~~IIII~~ ~~IIII~~ II	12
	Total	235

Note that we record the data in groups of five by recording four and then barring across for the fifth one. This is to make totalling each line of data easier.

The data recorded is for the full population for the given period of time. We could save time by sampling, i.e. perhaps recording the data for 10 minutes in any hour. The problem with this is that by reducing the time we could reduce the accuracy of the survey. In addition to this we could introduce bias into our survey because of our choice of 10 minutes, i.e. the first 10, the last 10, or 10 minutes out of the middle of the hour. The way in which we select and record our data is therefore of the utmost importance.

In order to work with the information gathered, i.e. to present and to analyse it, we group it into a smaller number of pieces of information to make it more manageable. These smaller pieces are called 'classes' and we look to form six to eight classes. The tally diagram shown on Fig. 4.1 has 12 individual sets of data so we could group them into six classes each of 2 hours' duration.

Fig. 4.2 shows the original data recorded in 2 hour classes. Notice how we now refer to the 'frequency' of the event taking place, in other words the frequency of a car passing over the bridge. We may also express the frequency by means of the relative frequency, this being the class frequency expressed as a percentage of the total frequency:

$$\text{the first class} = \frac{19}{235} \times \frac{100}{1} = 8.085\%$$

In this example the relative frequency has been rounded to the nearest 0.25%.

Figure 4.2

	Time (2 hour periods)	Frequency	Relative frequency
1	0600 – 0800	19	8.0
2	0800 – 1000	32	13.5
3	1000 – 1200	52	22.0
4	1200 – 1400	55	23.5
5	1400 – 1600	48	20.5
6	1600 – 1800	29	12.5
	Total	235	100.0

There is of course no reason why we should not combine our tally diagram and frequency table as shown in Fig. 4.3.

Figure 4.3

	Tally/frequency table Date:-		Frequency	Relative frequency
	Number of vehicles crossing bridge on Wednesday between 0600 and 1800 hours			
	Time 2 hour periods		Frequency	Relative frequency
1	0600–0800	⊞ ⊞ ⊞ IIII	19	8.0
2	0800–1000	⊞ ⊞ ⊞ ⊞ ⊞ ⊞ II	32	13.5
3	1000–1200	⊞ ⊞ ⊞ ⊞ ⊞ ⊞ ⊞ ⊞ ⊞ ⊞ II	52	22.0
4	1200–1400	⊞ ⊞ ⊞ ⊞ ⊞ ⊞ ⊞ ⊞ ⊞ ⊞ ⊞	55	23.5
5	1400–1600	⊞ ⊞ ⊞ ⊞ ⊞ ⊞ ⊞ ⊞ ⊞ III	48	20.5
6	1600–1800	⊞ ⊞ ⊞ ⊞ ⊞ IIII	29	12.5
		Totals	235	100.0

We have already mentioned that by grouping our data into classes we make it more manageable. Another reason is that some of the tallies may be so low that they are hardly worth recording as separate entities.

Grouping our data in this way gives us an immediate picture of the distribution of the event taking place, i.e. if we look at our tally diagram we can see the busiest period of use is between 1000 and 1500 hours.

Our data can also be presented pictorially in a number of different ways. The type of presentation depends on the use to which the data will be put.

Pictograms

With this form of presentation pictures will represent the data. The pictures used are related to the event and are given a numerical value, i.e. a matchstick figure to represent say 10 persons.

Using the data from our bridge crossing example, our pictogram could be drawn as shown in Fig. 4.4. Notice that incomplete vehicles are used to represent values less than 10.

Figure 4.4

	Time 2 hour periods	Vehicles crossing bridge 🚗 = 10 vehicles
1	0600–0800	🚗 🚗
2	0800–1000	🚗 🚗 🚗 🚗
3	1000–1200	🚗 🚗 🚗 🚗 🚗 🚗
4	1200–1400	🚗 🚗 🚗 🚗 🚗 🚗
5	1400–1600	🚗 🚗 🚗 🚗 🚗
6	1600–1800	🚗 🚗 🚗

Pie diagrams

These are sometimes called circle charts because we draw a circle to represent 100%. The 'pie' is then divided into wedges or sections to represent the data. The area of each wedge is proportional to the quantities being represented, i.e.

$$100\% = 360 \text{ degrees}$$

$$50\% = 360 \times \frac{50}{100} = 180 \text{ degrees}$$

$$22\% = 360 \times \frac{22}{100} = 79.2 \text{ degrees}$$

Fig. 4.5 shows a pie chart representing the data for our example.

Figure 4.5

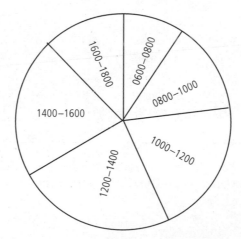

Bar charts

These can be represented in three different ways:

1. Horizontal – each bar is drawn horizontally to represent the percentage of the event occuring (Fig. 4.6).

Figure 4.6

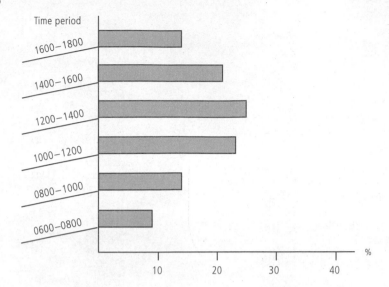

2. Vertical – each bar is drawn vertically to represent the percentage of the event occurring (Fig. 4.7).

Figure 4.7

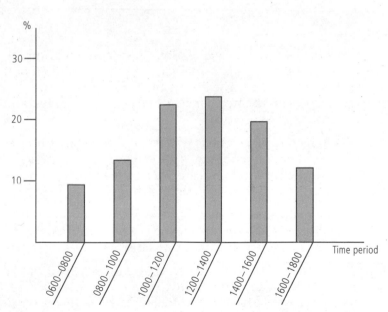

3. 100% barchart – in this representation the bar is usually drawn vertically to represent the 100% and is subdivided into sections showing the percentage of the event occurring. Each section is shaded differently to show the different percentages more clearly (Fig. 4.8).

Figure 4.8

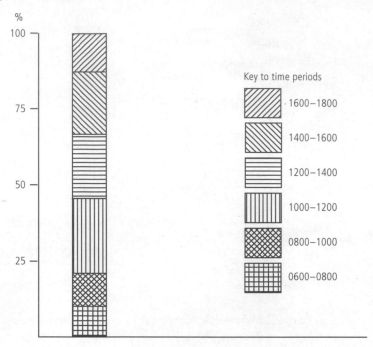

Histograms

This is generally considered to be one of the best representations of grouped data. It is important that the class intervals are the same and, if not, that they are adjusted to a true picture. The histogram is a series of vertical rectangles whose heights represent the frequencies occurring in each class and whose widths represent the class intervals.

The area of each rectangle, i.e. the frequency multiplied by the class interval, is proportional to the number of times the event occurred in each class. This means that when the class intervals are not all the same size, the height of the rectangles must be adjusted to maintain the proportionality.

When we draw a histogram we generally use the class mid-points on our scale, although, if preferred, we can use the class end-points. Fig. 4.9 shows the data from our example being drawn as a histogram using the midpoints.

Coming back to the question of unequal class intervals, let us take a simple example from our histogram. If we were to combine classes 0800–1000 and 1000–1200 then we would have a total frequency of 84. Now, since we have doubled the class length, we need to halve the combined heights to give us an area that is proportional to the other classes.

EXAMPLE 4.2

A building contractor decided to check on the periods of absenteeism amongst his staff and obtained the following data:

Figure 4.9

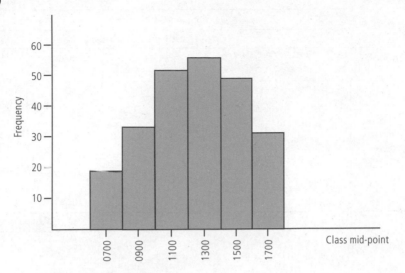

No. of days absent	Frequency
0	⊦⊦⊦⊦ III
1	⊦⊦⊦⊦ ⊦⊦⊦⊦ ⊦⊦⊦⊦ IIII
2	⊦⊦⊦⊦ ⊦⊦⊦⊦ ⊦⊦⊦⊦ I
3–5	⊦⊦⊦⊦ ⊦⊦⊦⊦ ⊦⊦⊦⊦ I
6–10	⊦⊦⊦⊦ IIII
11-15	⊦⊦⊦⊦ III
16–21	⊦⊦⊦⊦ ⊦⊦⊦⊦ ⊦⊦⊦⊦

Show this data on:

(a) a combined tally/frequency table
(b) pie chart
(c) horizontal barchart
(d) histogram.

(a) Combined tally/frequency table

This table is shown in Fig.4.10. The relative frequencies are calculated as shown below and rounded to one place of decimal. This introduced 0.1% of error which is small enough to ignore.

Relative frequency calculations:

0 days absent $\quad \dfrac{8}{91} \times 100 = 8.8\%$

1 day absent $\quad \dfrac{19}{91} \times 100 = 20.9\%$

2 days absent $\quad \dfrac{16}{91} \times 100 = 17.6\%$

Figure 4.10

Tally/frequency table					Frequency	Relative frequency
Staff absenteeism						
Days absent						
0	⊞	‖‖			8	8.8
1	⊞	⊞	⊞	‖‖‖	19	20.9
2	⊞	⊞	⊞	‖	16	17.6
3–5	⊞	⊞	⊞	‖	16	17.6
6–10	⊞	⊞	‖		9	9.9
11–15	⊞	‖‖‖			8	8.8
16–25	⊞	⊞	⊞		15	16.1
				Totals	91	100.0

3–5 days absent $\dfrac{16}{91} \times 100 = 17.6\%$

6–10 days absent $\dfrac{9}{91} \times 100 = 9.9\%$

11–15 days absent $\dfrac{8}{91} \times 100 = 8.8\%$

16–25 days absent $\dfrac{15}{91} \times 100 = 16.5\%$

(b) Pie chart

This is shown in Fig. 4.11 and the calculations to find the number of degrees for each class are shown below. To calculate the number of degrees, we express the relative frequency as a fractional part of the total and multiply by 360:

Figure 4.11

0 days absent $\qquad \dfrac{8.8}{100} \times 360 = 31.68°$

1 day absent $\qquad \dfrac{20.9}{100} \times 360 = 75.24°$

2 days absent $\qquad \dfrac{17.6}{100} \times 360 = 63.36°$

3–5 days absent $\qquad \dfrac{17.6}{100} \times 360 = 63.36°$

6–10 days absent $\qquad \dfrac{9.9}{100} \times 360 = 35.64°$

11–15 days absent $\qquad \dfrac{8.8}{100} \times 360 = 31.68°$

16–25 days absent $\qquad \dfrac{16.5}{100} \times 360 = 59.4°$

(c) Horizontal barchart

This is shown in Fig. 4.12. The relative frequencies are shown as horizontal bars equal in length to the percentages calculated in (b).

Figure 4.12

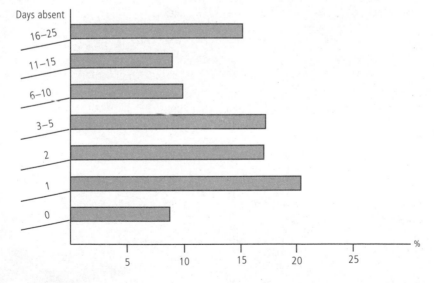

(d) Histogram

The first point to note is that the class intervals are not all the same size. We know from our previous example that if the intervals are the same size then the histogram is fairly straightforward to draw, but when we have unequal classes, it is a little bit more involved. You will recall that we stated earlier that histograms are based on areas representing each class, so for our example we must work out the area for each class so that we can produce an accurate histogram.

If we represent each day by 1 unit and each frequency occurrence by 1 unit then the total area for each occurrence is 1, i.e. if we had an interval of 25–33 days, there is a class interval equal to 9 days. Assuming we had a frequency of 27 in this class, the height of the histogram is 27/9 = 3. This gives us an area of 27 units square. Fig. 4.13 illustrates this histogram. Each box represents one frequency of the event occurring. The calculations for our example are:

Figure 4.13

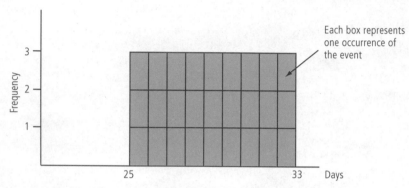

Each box represents one occurrence of the event

0 days absent gives a height of $\dfrac{8}{1} = 8$ units^2

1 day absent gives a height of $\dfrac{19}{1} = 19$ units^2

2 days absent gives a height of $\dfrac{16}{1} = 16$ units^2

3–5 days absent gives a height of $\dfrac{16}{3} = 5.33$ units^2

6–10 days absent gives a height of $\dfrac{9}{4} = 2.25$ units^2

11–15 days absent gives a height of $\dfrac{8}{5} = 1.6$ units^2

16–25 days absent gives a height of $\dfrac{15}{10} = 1.5$ units^2

The finished histogram is shown in Fig. 4.14.

See Section 4.4. for further application exercises. There are no anwers to the exercise due to the variety of scales etc. that can be selected.

Figure 4.14

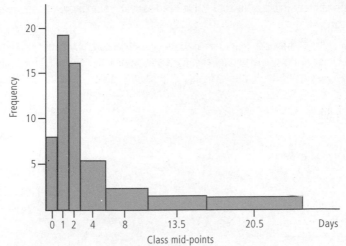

4.2 Measures of central tendency

We often refer to the average of a particular event occurring in such a way that it may or may not be meaningful. We talk about the average size of the male or female population, and the average weight of a person within a given age group. From these averages, certain information can be obtained, e.g. over the last 50 years the average height of the population has increased.

Sometimes these averages would appear to be rather silly, for instance when we talk about the average size of a family. Invariably this comes in fractional form such as 4.2 people, which raises the question of 0.2 people.

Often these averages can be a useful guide to manufacturers and producers of various commodities. If the local newspaper editor knows that 1 in 2 local inhabitants buys the newspaper then he has a fairly realistic idea how many papers to produce.

Because of the rather loose way in which we use the word average, statisticians prefer not to use it. Instead they prefer to use a measure of central tendency. The measure of central tendency tells us something about the centre of the distribution. The three most used measures of central tendency are the mean, the median and the mode.

The mean

The arithmetic mean, or as it is generally referred to, the mean is the mathematic term for average. It is calculated in the same way as the average, i.e. the sum of the values divided by their total number.

EXAMPLE 4.3

If we have five properties for sale at the following prices, £56 000, £61 000, £98 000, £58 000 and £55 000, then the total value is £328 000.

Therefore the arithmetic mean is the total value divided by the number of values, i.e. $328\,000 \div 5 = £65\,600$.

The arithmetic mean is the most important of the measures of central tendency and is generally stated in algebraic form. Since to find the mean we sum up a series of values, these summations are represented by the Greek letter upper case sigma which is written Σ. Now if we represent each value by x, i.e.

the first value is x_1

the second value is x_2

the third value is x_3

the fourth value is x_4

and the total number of values by n, then we can say that the mean is

$$\frac{x_1 + x_2 + x_3 + \ldots + x_n}{n} = \frac{\Sigma x}{n} = \bar{x}$$

$\bar{x}$ is the mean and referred to as 'x bar'.

Therefore the arithmetic mean can be represented by

$$\bar{x} = \frac{\Sigma x}{n}$$

More often than not when calculating the mean we will be using data in which the values occur a number of times. We may use this data either as ungrouped frequency distributions or grouped (i.e. put in classes) frequency distributions. Example 4.4 shows an ungrouped example.

EXAMPLE 4.4

Ungrouped data:

x	f	fx
13	6	78
14	3	42
15	7	105
16	6	96
18	14	252
19	9	171
20	11	220
21	7	147
	63	1111

$$\bar{x} = \frac{\Sigma fx}{n} = \frac{1111}{63} = 17.635$$

Although we have introduced a new symbol into our formula, f, nothing has changed. It is still the total of all the values, i.e. the value multiplied by its frequency f and each individual 'sub-total' added together, Σfx.

EXAMPLE 4.5

In preparing its end of the year report, a local council planning department finds that it has granted planning permission for the following number of domestic dwellings:

650 four bedroom detached houses

986 three bedroom detached houses

321 four bedroom semi-detached houses

423 three bedroom semi-detached houses

526 four bedroom bungalows

872 three bedroom bungalows

606 four bedroom bungalows

Find the arithmetic mean for:

(a) the total developments
(b) houses only
(c) three bedroom dwellings.

(a) To find the mean of all the dwellings we need to know the total number of dwellings and how many different types.

Types of dwelling	x	f	fx
4 bedroom detached houses	1	650	650
3 bedroom detached houses	1	986	986
4 bedroom semi-detached houses	1	321	321
3 bedroom semi-detached houses	1	423	423
4 bedroom bungalows	1	526	526
3 bedroom bungalows	1	872	872
2 bedroom bungalows	1	606	606
	$n = 7$		$\Sigma fx = 4384$

$$\bar{x} = \frac{\Sigma fx}{n} = \frac{4384}{7} = 626.29 \text{ dwellings}$$

(b) To find the mean we need to know the total number of houses and how many different types.

Types of houses	x	f	fx
4 bedroom detached housess	1	650	650
3 bedroom detached houses	1	986	986
4 bedroom semi-detached houses	1	321	321
3 bedroom semi-detached houses	1	423	423
	$n = 4$		$\Sigma fx = 2380$

$$\bar{x} = \frac{\Sigma fx}{n} = \frac{2380}{4} = 595 \text{ houses}$$

(c) Here we need to know the total number of three bedroom dwellings and how many different types.

Three bedrooms	x	f	fx
Detached houses	1	986	986
Semi-detached houses	1	423	423
Bungalows	1	872	872
	n = 3		Σ fx = 2281

$$\bar{x} = \frac{\Sigma fx}{n} = \frac{2281}{3} = 760.33 \text{ three bedroom dwellings}$$

Calculating the mean from grouped data is basically the same as for ungrouped data. Remember that the purpose of grouping is to reduce the number of different variables to a manageable size, i.e. if we have say 20 different values we put them into six to eight classes and work around the mid-point. The mid-point value is multiplied by the total frequency of that class to give *fx*. The following example shows the procedure for calculating the mean from a grouped frequency distribution.

EXAMPLE 4.6

Class interval mid-point frequency:

	x	f	fx
5–19	12	8	96
20–34	27	17	459
35–49	42	29	1218
50–64	57	41	2337
65–79	72	27	1944
80–94	87	16	1392
		n = 138	Σ fx = 7446

$$\bar{x} = \frac{\Sigma fx}{n} = \frac{7446}{138} = 53.96$$

EXAMPLE 4.7

A brick manufacturer wishes to know the mean length of the bricks that he produces. Over a period of time the following data was obtained. Calculate the arithmetic mean length of the bricks.

Length (mm)	Frequency
205	2
205.5	1
206	4
206.5	3
207	9
207.5	4
208	7
208.5	8
209	6
209.5	7
210	9
210.5	9
211	12
211.5	16
212	21
212.5	32
213	48
213.5	61
214	72
214.5	78
215	87
215.5	72
216	59
216.5	36

Here we have a wide range of values for the length, so the data would be much easier grouped into classes. Since there are 24 different values, it would lend itself to eight classes, each class spanning three values.

Class interval	mid-point (x)	frequency (f)	fx
205–206	205.5	7	1438.5
206.5–207.5	207	16	3312
208–209	208.5	21	4378.5
209.5–210.5	210	25	5250
211–212	211.5	49	10363.5
212.5–213.5	213	141	30033
214–215	214.5	237	50836.5
215.5–216.5	216	167	36072
		$n = 663$	$\Sigma fx = 141684$

$$\bar{x} = \frac{\Sigma fx}{n} = \frac{141684}{663} = 213.7$$

The mean length of the bricks produced is 213.7 mm.

The median

The median is the value that occurs at the middle of a set of values. In order to find the value of the median, the set of values is ranked, i.e. placed in order of value generally with the lowest to the left and increasing to the right. If the set contains an odd number of values then the middle term is selected as being the median of the set.

EXAMPLE 4.8

Find the median of the following set of numbers: 8, 11, 15, 9, 17, 12, 10.

Ranking the numbers we get: 8, 9, 10, 11, 12, 15, 17. Since there is an odd number of values we take middle value as the median, i.e. 11.

When there is an even number of values in the set then to find the median we take the mean of the two middle values.

EXAMPLE 4.9

Find the median of the following set of numbers: 61, 80, 75, 59, 68, 72, 90, 89.

Ranking the numbers we get: 59, 61, 68, 72, 75, 80, 89, 90. The middle two values are 72 and 75.

Therefore the median is equal to 72 + 75 = 73.5.

The main advantage of using the median as a measure of central tendency is that extreme values in a set do not have a bias effect. If we considered the following set of numbers, 8, 10, 11, 15, 63, then the median is 11. Even if the extreme value of 63 had been greater, the median would still be 11 and representative of the set.

The mode

This is another measure of central tendency where extreme values occur and one which is easily determined since it is found by observation rather than calculation. The mode is the value that occurs the most frequently. Consider the following set of numbers:

7, 8, 10, 14, 15, 12, 8, 9, 11, 10, 12, 13, 10, 12

The value 10 is the mode since it has the highest frequency.

For grouped data the mode is the class with the highest frequency and is often referred to as the modal class.

EXAMPLE 4.10

A local authority decides to undertake to find the size of families in its area as part of its preparation for its housing policy and obtained the following results:

No. of children	0	1	2	3	4	5
No. of families	47	180	296	102	61	14

Calculate the mean, median and mode of the families.

The mean is equal to the total number of children divided by the number of families surveyed:

$$\bar{x} = \frac{\text{total no. of children}}{\text{total no. of families}}$$

$$= \frac{(0 \times 47) + (1 \times 180) + (2 \times 296) + (3 \times 102) + (4 \times 61) + (5 \times 14)}{47 + 180 + 296 + 102 + 61 + 14}$$

$$= \frac{180 + 592 + 306 + 244 + 70}{700}$$

$$= \frac{1392}{700}$$

$$= 1.99 \text{ children}$$

The median can be found by taking the middle value of the total number of children:

$$1392 \div 2 = 696$$

Therefore the median is the mean of 696 and 697, that is 696.5. We now need to know where this value occurs. We can find this simply by locating the set in which this value occurs. If we add the values starting at zero children, we find:

$$0 \times 47 = 0$$

$$1 \times 180 = 180$$

$$2 \times 296 = 592$$

$$592 \times 180 = 772$$

The 696.5 value occurs between 180 and 772, and therefore the median value is 2 children.

The largest group is that of families with 2 children; therefore the mode is 2 children.

From our results we can see that mathematically the mean is more accurate but practically the results are all the same since the common real value is 2 children.

See Section 4.4 for further application exercises.

4.3 Frequency distributions

We have already come across the use of frequency distributions for pictorial representation in Section 4.1. We are now going to look at them in a little more detail.

It would be useful to be more precise about the data we are using. In the main I have used such terminology as values etc. when referring to our raw data, whether this has been height, lengths or items. To be more correct I should refer to these quantities as variables.

Variables can be of two types:

1. *Discontinuous or discrete variables.* These are variables that take on a complete value no matter how small and have the quality of being able to be counted, i.e. the number of people in a room, the number of bricks bought or the number of nails in a kilogram.

2. *Continuous variables.* These variables all have values within certain limits but arguably no finite value and so are said to be continuous since they are defined by measurement. In this case we decide on a value within given limits. Examples of continuous variables are:
   (a) time: how accurately are we going to measure this ongoing variable – hours, minutes, seconds, fractions of a second?
   (b) weight and length both follow the same sort of argument as above; we determine the accuracy but whatever we decide it could always be smaller or larger.

We have already met with the idea of frequency tables and what we need to do now is to carry these ideas further forward with one or two examples.

EXAMPLE 4.11

The lengths of 100 facing bricks whose nominal length is 215 mm were measured as part of a quality control. Check the following results obtained:

215	214	214	216	215	213	214	215	215	216
213	215	212	215	218	214	215	213	217	215
212	214	213	214	215	215	215	216	213	215
213	215	216	213	214	214	212	214	213	216
218	215	215	214	219	215	214	216	215	214
214	213	214	215	213	215	215	216	217	213
216	217	215	215	216	214	215	215	215	214
217	216	215	214	215	215	214	215	216	217
216	215	214	213	217	215	218	216	214	215
215	214	215	216	215	215	214	215	217	216

(a) Prepare a frequency distribution table.

By observation we can see that the range of our variants is from 211 to 219. This will give us nine variables using the 1 mm interval (see Fig. 4.15).

From this table we can see that the length 215 mm occurs 37 times and 219 mm only once and so on.

(b) Draw a histogram of the frequency distribution. Fig. 4.16 shows the histogram. Remember that with a histogram it is the areas that represent the frequencies. Since in this example there are no unequal classes, the width of the histogram bars will all be the same.

By joining the mid-points at the top of the rectangles we get a frequency polygon. This again helps to show the spread of the data.

Figure 4.15

Length (mm)	Frequency distribution table 100 brick lengths	Frequency
211	I	1
212	III	3
213	⊞⊞ ⊞⊞ I	11
214	⊞⊞ ⊞⊞ ⊞⊞ ⊞⊞ II	22
215	⊞⊞ ⊞⊞ ⊞⊞ ⊞⊞ ⊞⊞ ⊞⊞ ⊞⊞ II	37
216	⊞⊞ ⊞⊞ ⊞⊞	15
217	⊞⊞ II	7
218	III	3
219	I	1
	Total	100

Figure 4.16

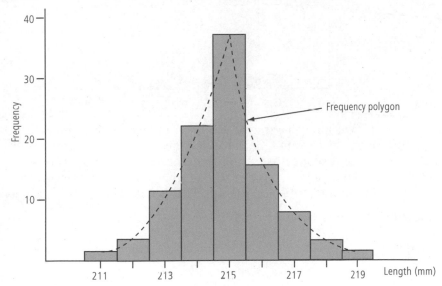

(c) Arrange the data into classes and draw a histogram of the grouped data.

When dealing with large amounts of data, the grouping of the data into classes makes it easier to work with. The first step is to arrange the data into a grouped frequency distribution table. This is shown in Fig. 4.17. To draw the histogram we have to decide on the dividing point between the classes, in other words the accuracy of the mid-point. I have decided to work to one place of decimals so that the first class is 210.5 to 212.5, the second class is 212.5 to 214.5 and so on. These measurements are known as the upper and lower class boundaries and the difference between is called the class width, i.e.

class width = upper class boundary − lower class boundary

= 212.5 − 210.5

= 2 mm

The histogram can now be drawn using the class mid-points (see Fig. 4.18).

Figure 4.17

Grouped frequency distribution table	
Class	Frequency
210.5–212.5	4
212.5–214.5	33
214.5–216.5	52
216.5–218.5	10
218.5–220.5	1
Total	100

Figure 4.18

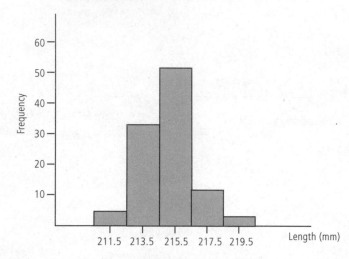

In our example we showed a frequency polygon by drawing through the top mid-point on our histogram (Fig. 4.16). Frequency polygons can take on a number of different shapes. Many are symmetrical about a central point, the most common being a bell shape as shown in Fig. 4.19 which tends to illustrate a common distribution of data. This is the main one as far as we are

Figure 4.19

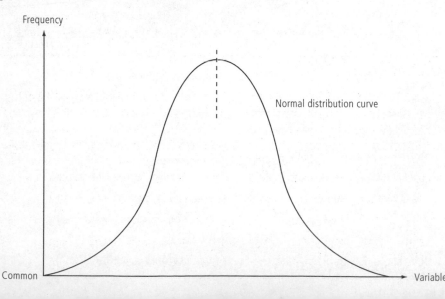

concerned and it is called a normal distribution curve. Two other fairly common distribution curves are shown in Figs 4.20 and 4.21.

Figure 4.20

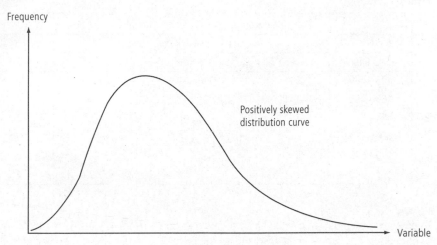

Positively skewed distribution curve

Figure 4.21

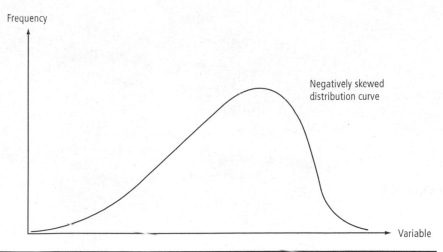

Negatively skewed distribution curve

EXAMPLE 4.12

A random selection of 60 steel rods were tested to destruction. The results of tests showing the maximum stress applied to each bar are shown below. Construct a frequency distribution table and histogram of the data which is in N/mm².

510	500	495	500	480	495	480	475	490	495
525	505	510	495	500	520	480	485	475	490
485	500	505	500	510	505	500	495	515	520
525	510	495	500	505	485	500	500	490	475
485	525	490	505	500	500	505	495	500	500
475	495	525	515	510	500	500	505	495	510

The first step is to prepare a frequency distribution table. By observation we can see that the range of values is 475 to 525 N/mm² which I shall group into six classes (see Fig. 4.22).

Figure 4.22

Class limits	Grouped frequency distribution table		Frequency
	Tensile test result on mild steel		
475–480	~~IIII~~ III		8
485–490	~~IIII~~ III		8
495–500	~~IIII~~ ~~IIII~~ ~~IIII~~ ~~IIII~~ III		23
505–510	~~IIII~~ ~~IIII~~ III		13
515–520	IIII		4
525–530	IIII		4
		Total	60

To draw the histogram we need to find the class mid-points. From 475 to 480 there are six values, so our mid-point is between the third and fourth, i.e. 477.5. In this way we can find all the mid-points:

475–480   mid-point   477.5

485–490   mid-point   487.5

495–500   mid-point   497.5

505–510   mid-point   507.5

515–520   mid-point   517.5

525–530   mid-point   527.5

We can now construct our histogram (see Fig. 4.23).

Figure 4.23

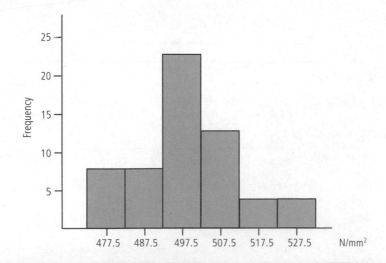

See Section 4.4 for further application problems.

4.4 Standard deviation

We know that the common measures of central tendency are the mean, mode and median. The mean is the one that is generally preferred. Having established our measure of central tendency, the next important fact that we will probably need to establish is how much the data deviates from this measure. To measure this dispersion we calculate the standard deviation. This measure of dispersion or scatter of data is represented by the lower case greek letter sigma, σ.

If the value of the standard deviation is small, this suggests that the value of the variant is very close to the measure of central tendency. For example, if we assume that the mean strength of concrete blocks is $14\,N/mm^2$ and the standard deviation is $0.1\,N/mm^2$ then to a statistician this indicates that 95% of the blocks will be within the strength range of 13.8 to $14.2\,N/mm^2$, whereas if the standard deviation is $0.3\,N/mm^2$ then 95% of the blocks will be within the strength range of 13.4 to $14.6\,N/mm^2$.

To find the standard deviation of a set of variants is fairly straightforward providing we follow a simple procedure.

To calculate the standard deviation (σ):

Step 1. Calculate the arithemetic mean $\bar{x}$.

Step 2. Find the deviation of each variant, i.e.

$$(x_1 - \bar{x}), (x_2 - \bar{x}), (x_3 - \bar{x}) \ldots (x_n - \bar{x})$$

Step 3. Square each deviation: $(x - \bar{x})^2$; this removes any negative signs.

Step 4. Sum all the squares, i.e.

$$(x_1 - \bar{x})^2 + (x_2 - \bar{x})^2 + (x_3 - \bar{x})^2 \ldots (x_n - \bar{x})^2 = \Sigma(x - \bar{x})^2$$

Step 5. Divide by the number of variants, i.e.

$$\frac{\Sigma(x - \bar{x})^2}{n}$$

to find the mean square value.

Step 6. Take the square root of the mean squares to reverse the squaring in step 3, i.e. for ungrouped data

$$\text{standard deviation } \sigma = \sqrt{\frac{\Sigma(x - \bar{x})^2}{n}}$$

and for grouped data

$$\text{standard deviation } \sigma = \frac{\Sigma f(x - \bar{x})^2}{\Sigma f}$$

EXAMPLE 4.13

Calculate the mean and standard deviation of the following numbers:

21, 25, 23, 18, 19, 22, 20, 24, 17

The mean as we know is the sum of all the numbers divided by the number of variants:

$$\bar{x} = \frac{\Sigma x}{n}$$

$$= \frac{21 + 25 + 23 + 18 + 19 + 22 + 20 + 24 + 17}{9}$$

$$= \frac{189}{9}$$

$$= 21$$

To calculate the standard deviation we follow the steps listed previously.

Step 1. $\bar{x} = 21$

Step 2. To find the deviation of each variant $(x_n - \bar{x})$:

$21 - 21 = 0$

$25 - 21 = 4$

$23 - 21 = 2$

$18 - 21 = -3$

$19 - 21 = -2$

$22 - 21 = 1$

$20 - 21 = -1$

$24 - 21 = 3$

$17 - 21 = -4$

Step 3. Square each deviation, $(x_n - \bar{x}^2)$:

$0^2 = 0$

$4^2 = 16$

$2^2 = 4$

$-3^2 = 9$

$-2^2 = 4$

$1^2 = 1$

$-1^2 = 1$

$3^2 = 9$

$-4^2 = 16$

Step 4. Sum all the squares $\Sigma(x - \bar{x})^2$:

$$0 + 16 + 4 + 9 + 4 + 1 + 1 + 9 + 16 = 60$$

Step 5. Divide by the number of variants $\dfrac{\Sigma(x - \bar{x})^2}{n}$

$$\frac{60}{9} = 6.67$$

Step 6. Take the root of $\dfrac{\Sigma(x - \bar{x})^2}{n}$ to give σ:

$$\sigma = \sqrt{\frac{\Sigma(x - \bar{x})^2}{n}}$$

$$= \sqrt{6.67}$$

$$= 2.58$$

As you can see it is a step by step by step procedure which, if we stick to it, will allow us to calculate the standard deviation with comparative ease. Now that we have tried the procedure we can speed it up by recording and working on our data in tabulated form.

EXAMPLE 4.14

Sixteen concrete cubes were tested for compressive strength and the failing loads in kN were:

640	680	685	700	680	670	675	680
660	690	685	690	685	685	675	680

Calculate the mean failing load and the standard deviation of the data.

If we tabulate our data it will make working the different steps much easier:

Failing load (x)	Deviation ($x - \bar{x}$)	Deviation^2 ($x - \bar{x})^2$
640	−38.75	1501.5625
660	−18.75	351.5625
670	−8.75	76.5625
675	−3.75	14.0625
675	−3.75	14.0625
680	1.25	1.5625
680	1.25	1.5625

680	1.25	1.5625
680	1.25	1.5625
685	6.25	39.0625
685	6.25	39.0625
685	6.25	39.0625
685	6.25	39.0625
690	11.25	126.5625
690	11.25	126.5625
700	21.25	451.5625
$\Sigma x = 10\,860$		$\Sigma = 2825$

$$\bar{x} = \frac{\Sigma x}{n} \qquad\qquad \sigma = \sqrt{\frac{\Sigma(x - \bar{x})^2}{n}}$$

$$\bar{x} = \frac{10\,860}{16} \qquad\qquad \sigma = \sqrt{\frac{2825}{16}}$$

The mean $\bar{x} = 678.75$ kN $\qquad\qquad \sigma = \sqrt{176.5625}$ kN

and the standard deviation $\sigma = 13.288$ kN

EXAMPLE 4.15

The weekly output for 10 consecutive weeks from a cement producing works in tonnes is

115	100	120	116	118	116	117	118	115	118

Calculate the standard deviation.

Output (tonnes) (x)	Deviation $x - \bar{x}$	Deviation^2 $(x - \bar{x})^2$
100	−15.3	234.09
115	−0.3	0.09
115	−0.3	0.09
116	0.7	0.49
116	0.7	0.49
117	1.7	2.89
118	2.7	7.29
118	2.7	7.29
118	2.7	7.29
120	4.7	22.09

$$\Sigma x = 1153 \qquad\qquad\qquad \Sigma = 282.91$$

$$\bar{x} = \frac{\Sigma x}{n} \qquad\qquad\qquad \sigma = \sqrt{\frac{\Sigma(x - \bar{x})^2}{n}}$$

$$\bar{x} = \frac{1153}{10} \qquad\qquad\qquad \sigma = \sqrt{\frac{282.9}{10}}$$

$$\bar{x} = 115.3 \text{ tonnes} \qquad\qquad \sigma = \sqrt{28.29} \text{ tonnes}$$

$$\sigma = 5.31 \text{ tonnes}$$

For grouped data we follow the same procedure but this time we have to worked around the mid-points of each class.

EXAMPLE 4.16

The following data represents the results of a series of tensile tests on 100 steel specimens. Calculate the standard duration of the results.

Class limits (N/mm²)	Frequency
600–604	2
605–609	7
610–614	19
615–619	28
620–624	25
625–629	14
630–634	4
635–639	1

As with ungrouped data it is a good idea to tabulate the results to assist in the calculations. Our table will need one or two extra columns to allow for the mid-point etc.

We will need to determine the mid-point of each class, i.e.

600–604   the mid-point is 602

605–609   the mid-point is 607

610–614   the mid-point is 612

So, having entered our classes in our table we now enter the mid-points followed by their frequencies so that we can calculate the product. We now proceed as before, finding the deviation, squaring it and finally multiplying by the frequency.

Class limits	Mid-point x	Frequency f	Product fx	Deviation $(x-\bar{x})$	$(x-\bar{x})^2$	$f(x-\bar{x})^2$
600–604	602	2	1204	−16.5	272.25	544.5
605–609	607	7	4249	−11.5	132.25	925.75
610–614	612	19	11628	−6.5	42.25	802.75
615–619	617	28	17276	−1.5	2.25	63.0
620–624	622	25	15550	3.5	12.25	306.25
625–629	627	14	8778	8.5	72.25	1011.5
630–634	632	4	2528	13.5	182.25	729.0
635–639	637	1	637	18.5	342.25	342.25
		$\Sigma\,100$	$\Sigma\,61850$			$\Sigma\,4725$

$$\text{Mean } \bar{x} = \frac{\Sigma fx}{\Sigma f} = \frac{61850}{100} = 618.5\,\text{N/mm}^2$$

$$\text{Standard deviation } \sigma = \sqrt{\frac{\Sigma f(x-\bar{x})^2}{\Sigma f}}$$

$$= \sqrt{\frac{4725}{100}}$$

$$= 6.87\,\text{N/mm}^2$$

EXAMPLE 4.17

The following data represents the number of bricks laid by 50 bricklayers per day. Calculate the mean and standard deviation.

300	310	290	295	250	240	250	260	265	305
275	280	270	240	290	310	325	295	305	280
260	275	280	275	305	300	275	275	280	275
250	260	275	270	290	280	310	325	320	315
290	275	270	265	270	295	300	280	270	265

We need to order our data into classes and produce a tally frequency table. By observation we can see that the range of the data is from 240 to 320, giving a spread of 80. I shall say 90 and have six classes with a spread of 15.

No. of bricks per hour		Frequency
240–254	⊬Ⲧ	5
255–269	⊬Ⲧ I	6
270–284	⊬Ⲧ ⊬Ⲧ ⊬Ⲧ III	18
285–299	⊬Ⲧ III	8
300–314	⊬Ⲧ IIII	9
315–329	IIII	4

Class limits	Mid-point x	Frequency f	Product fx	Deviation $(x - \bar{x})$	$(x - \bar{x})^2$	$f(x - \bar{x})^2$
240–254	248	5	1240	−36.6	1339.56	6697.8
255–269	263	6	1578	−21.6	466.56	2799.36
270–284	278	18	5004	−6.6	43.56	784.08
285–299	293	8	2344	8.4	70.56	564.48
300–314	308	9	2772	23.4	547.56	4928.04
315–329	323	4	1292	38.4	1474.56	5898.24
		50	Σ 14 230			21 672.00

$$\bar{x} = \frac{\Sigma fx}{\Sigma f} = \frac{14\,230}{50}$$

$$= 284.6 \text{ bricks}$$

$$\sigma = \sqrt{\frac{\Sigma f(x - \bar{x})^2}{\Sigma f}}$$

$$= \sqrt{\frac{21\,672}{50}}$$

$$= 20.82 \text{ bricks}$$

See Section 4.4. for further application problems.

Finally let us have another look at the normal distribution curve we referred to earlier. The area under any distribution curve is equal to the population. A normal distribution curve has certain properties regarding this area. If we measure one standard deviation either side of the measure of central tendency, i.e. the mean, then the area under the curve is equal to 68% of the population. We say ± 2 standard deviations is equal to 68% (see Fig. 4.24).

Two standard deviations either side of the mean gives an area under the curve equal to 95%, i.e. ± 2 standard deviations = 95% (Fig. 4.24).

Three standard deviations either side of the mean covers 100%, i.e. ± 3 standard deviations = 100%.

Mathematically this is not strictly true, but it is so close that for practical purposes it is regarded as true.

It is for these reasons that, from a quality point of view, the smaller the standard deviation the more of our products are likely to be acceptable.

Figure 4.24

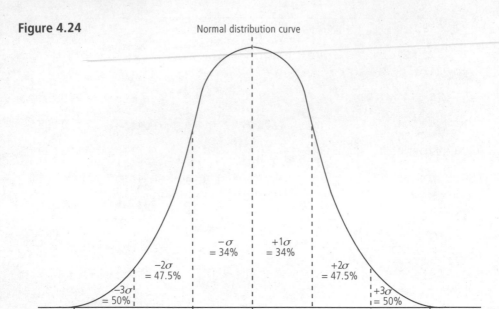

Normal distribution curve

4.5 Application exercises

Pictorial representation

1 A brick manufacturer's output of bricks per week is as follows:

brick type A = 120 000

type B = 190 000

type C = 70 000

type D = 220 000

Illustrate this data as

(a) a pie chart
(b) a horizontal barchart
(c) a 100% vertical barchart.

2 A quality inspector at a window-making factory checked 200 sashes for general defects and obtained the following results:

oversized	15
undersized	23

out of square   12

damaged   20

good condition   130

Represent this data as

(a) a pictogram
(b) a pie chart
(c) a histogram.

3 The number of screws in 100 boxes selected at random from a production
   line were counted and the following results obtained:

100	98	97	100	101	100	100	100	100	102
98	100	100	98	98	100	97	99	101	100
99	100	99	100	100	98	101	100	100	100
100	100	99	101	98	99	100	97	100	100
100	97	100	100	100	100	97	100	98	99
100	99	101	100	99	97	98	97	102	100
100	100	100	100	101	100	100	101	100	100
97	100	102	99	100	97	102	100	100	99
100	98	100	100	100	99	100	100	101	100
99	100	100	100	100	98	100	97	100	98

Produce a tally/frequency diagram and a vertical barchart.

4 The crushing strength of 100 insulation blocks under test were as follows:

Crushing strength	5 N/mm²	6 N/mm²	7 N/mm²	8 N/mm²
No. of blocks	5	12	64	19

Show these results as a pie chart and a histogram.

5 The results of test for the thermal conductivity (W/mK) on 100 insulation
   blocks were as follows:

W/mK	No. of blocks
0.18	8
0.19	13
0.20	58
0.21	15
0.22	6

Show these results in the form of a histogram.

Measures of central tendency

1 Calculate the mean, median and mode of the following values:

27, 31, 32, 28, 30, 31,31, 29, 28, 29, 31, 33, 31

2 In an examination the following results were obtained:

% mark	30	40	50	60	70	80	90	100
No. of students	4	7	21	33	45	39	12	1

Calculate the arithmetic mean of the results.

3 A contractor builds the following walls for the given prices. Calculate the mean cost per square metre.

Area of walls (m²)	60	78	69	80
Cost of walls (£)	1500	2168.4	1718.1	2292

4 Three hundred trees were felled and the girth and length measured and recorded below:

Girth (mm)	500–640	650–790	800–940	950–1090
Length (m)	12	14	16	18
No. of trees	90	120	60	30

(a) Calculate the arithmetic mean, median and modal class of the girths and lengths.

(b) Using the arithmetic means calculate the volume.

5 A speculator purchases shares in the following construction company as shown. The stockbroker's commission for the transactions are also shown.

Company A   500 shares @ 175p each   commission = £12.50

Company B   750 shares @ 190p each   commission = £14.75

Company C   1000 shares @ 120p each   commission = £16.00

Company D   400 shares @ 250p each   commission = £15.75

Company E   500 shares @ 210p each   commission = £15.00

Calculate the mean costs of the shares including the commission.

Frequency distributions

1 The crushing strength for concrete cubes is given in the following frequency distribution table. Draw the histogram and show the frequency polygon for the data.

Crushing strength (N/mm²)	Frequency
19.42–19.44	11
19.45–19.47	41
19.48–19.50	84
19.51–19.53	45
19.54–19.56	9

2 Density tests on 30 timber specimens produce the following data:

Density (kg/m³)	570	580	590	600	610	620	630
Frequency	2	3	5	9	6	2	3

Draw the histogram and calculate the arithmetic mean.

3 The table below shows the percentage moisture constant of 200 × 50 mm timbers to be used as first floor joists. Construct a histogram and frequency distribution table.

18.5	18.6	19.0	19.1	18.8	17.9	17.8	17.9	18.0	18.1
17.7	18.2	18.1	19.0	18.9	18.7	17.9	18.1	18.1	18.2
18.6	18.5	17.9	18.1	19.0	18.9	18.6	18.6	18.7	18.5
17.9	19.1	18.0	18.1	18.2	18.2	18.5	18.4	17.8	19.0
18.9	18.7	18.6	18.6	18.6	18.4	18.0	18.1	18.1	17.9
18.1	18.0	18.1	18.7	19.1	18.2	17.9	18.4	18.3	18.4

4 The crushing strengths of 60 insulation blocks were determined and the results are recorded below. Draw the histogram and the frequency polygon.

N/mm²	6.6	6.8	7.0	7.2	7.4	7.6	7.8
Frequency	1	5	14	20	15	3	2

There are no answers to the problems in this section since there are a variety of scales that can be used.

Standard deviation

1 Calculate the mean and standard deviation of the following set of numbers:

18, 19, 16, 21, 23, 20, 22, 30

2 The annual population of an overspill village to a large city was recorded over a 12 year period. Calculate the mean and standard deviation of the population.

Year	1	2	3	4	5	6	7	8	9	10	11	12
Population (1000s)	3.2	3.3	3.1	3.3	3.4	3.6	3.3	3.2	3.3	3.4	3.4	3.5

3 The percentage of lead contained in 20 paint samples analysed in one week gave the following data:

2.41	2.39	2.38	2.40	2.41	2.42	2.39
2.40	2.39	2.41	2.38	2.37	2.41	2.40
2.39	2.37	2.41	2.41	2.39	2.37	

Calculate the mean and standard deviations.

4 The times taken for a man to assemble 10 window frames on a mass production assembly are (in minutes):

8.5 9.2 8.7 9.1 10.2 9.7 9.8 9.1 10.0 9.7

Calculate the mean time and the standard deviation.

5 In a series of tests on plastic components, the following results were obtained:

Class limits (N)	Frequency
220–224	2
225–229	3
230–234	6
235–239	16
240–244	20
245–249	17
250–254	5
255–259	2
260–264	1

Calculate the mean and standard deviation.

6 In a series of tests on 30 samples of copper the moduli of elasticity were:

106.4	120	105.2	104.8	116.5	112.1
110.2	111	108.9	109.2	114.3	113.4
119.6	109.7	117.1	115.9	112.6	100.7
114.7	108.7	110.2	112.3	109.5	108.1
110	111.3	110.1	112.4	108.9	111.5

Calculate the mean and standard deviation.

7 The frequency distribution for the thickness of slates is shown below. Calculate the mean and standard deviation.

Thickness (mm)	Frequency
4.0–4.2	13
4.3–4.5	57
4.6–4.8	82
4.9–5.1	64
5.2–5.4	24

8 The thermal conductivities of 30 bricks were measured and the following results obtained:

0.73	1.04	0.81	0.96	0.76	1.02
1.05	1.14	0.92	1.03	0.77	0.88
1.11	1.17	1.21	1.36	1.15	1.23
1.28	1.19	1.17	1.08	0.91	0.88
1.02	0.89	1.07	1.05	0.95	1.04

Calculate the mean and standard deviation.

4.6 Answers to Section 4.5

Measures of central tendency

1 Mean = 30.08

  Median = 31

  Mode = 31

2 Arithmetic mean = 67.16%

3 Mean cost/m^2 = £26.75

4 Girth: a.m. = 737.5 mm

      median = 720.6 mm

      modal class = 650–790

  Length: a.m. = 14.2 m

      median = 14 m

      mode = 14 m

  Volume = 184.38 m^3

5 178.5p

Standard deviation

1 $\bar{x} = 21.125$ $\sigma = 3.95$

2 $\bar{x} = 3.33$ $\sigma = 0.13$

3 $\bar{x} = 2.395\%$ $\sigma = 0.015\%$

4 $\bar{x} = 9.4$ min $\sigma = 0.535$ min

5 $\bar{x} = 241.44$ $\sigma = 7.66$

6 $\bar{x} = 111.177$ $\sigma = 4.15$

7 $\bar{x} = 4.74$ mm $\sigma = 0.32$

8 $\bar{x} = 1.029$ $\sigma = 0.156$

Index